The Patrick Moore Practical Astronomy Series

More information about this series at http://www.springer.com/series/3192

Remote Observatories for Amateur Astronomers

Using High-Powered Telescopes from Home

Gerald R. Hubbell
Richard J. Williams
Linda M. Billard

 Springer

Gerald R. Hubbell
Locust Grove, VA, USA

Richard J. Williams
Markleeville, CA, USA

Linda M. Billard
Fredericksburg, VA, USA

ISSN 1431-9756   ISSN 2197-6562   (electronic)
The Patrick Moore Practical Astronomy Series
ISBN 978-3-319-21905-9   ISBN 978-3-319-21906-6   (eBook)
DOI 10.1007/978-3-319-21906-6

Library of Congress Control Number: 2015948254

Springer Cham Heidelberg New York Dordrecht London

Cover Illustration by Rachel Konopa

Printed on acid-free paper

Springer International Publishing AG Switzerland is part of Springer Science+Business Media (www.springer.com)

Foreword I

by Scott W. Roberts

The purpose of an astronomical observatory is to contain the instrumentation that allows astronomers to observe and make measurements of the cosmos. They are often built—sometimes at great cost—in places that are the most optimum for the best observations, with today's observatories located both on land and in space for the use of professional and amateur astronomers. At the very least, observatories are built to make repeatable observations routine and convenient. In fact, some specialized observations can only be made from particular observatories with very specific equipment and/or locations.

Upon reading this book, you might feel that the advanced technologies we professional and amateur astronomers use every day to make and record scientific observations have reached new, unprecedented heights. And yet most certainly, many more peaks of innovation and discovery will be conquered at a pace that perhaps will leave even us—who are now quite accustomed to the breakneck pace of technology and our nearly continuous new discoveries about the universe we live in—uncomfortably numb.

We know that astronomical observatories were developed at least 10,000 years ago. And we could perhaps infer from what we are learning about humanity that the development leading to the building of Zorats Karer (in present-day Armenia), built in the sixth millennium BC, and Stonehenge, built in the second millennium BC more than 2600 miles away, that there are probably other ancient observatories yet to be discovered.

You could argue that without observatories, humanity might not have survived at all. Without accurate determinations of the changes of the seasons, accurate calendars to segment our orbits around the Sun, and accuracy of segmenting a single day, humanity might have been utterly lost and defenseless against the ravages of nature and perhaps even each other. The permanent placement of astronomical instrumentation in observatories allows repeatable and reliable measurements of the Sun, the planets, and the stars. Observatories are centers of higher learning, providing a stage for scientists from which they can disseminate the collected data and what has been learned from it to the scientific community and to the public at large.

Indeed, observatories are the vehicles that we developed to help us navigate and advance civilization and are as important to humanity as any invention that we have ever created. And today, all of the elements required to make a cutting-edge, high-technology, computer-controlled telescope and observatory system operated from anywhere on the planet through the Internet are within reach of dedicated amateur astronomers. What has been missing is a comprehensive guide for putting it all together.

Remote Observatories for Amateur Astronomers: Using High-Powered Telescopes from Home, written by Jerry Hubbell, Rich Williams, and Linda Billard, is a unique contribution centering on computer-controlled private observatories owned by amateur astronomers and commercialized professional–amateur observatories where observing time to collect data can be purchased. The methodical approach to operating a modern computer-controlled observatory and the discipline of critical thinking in pursuit of developing an astronomical imaging system (AIS) make this book a perfect companion to Hubbell's first book, *Scientific Astrophotography: How Amateurs Can Generate and Use Professional Imaging Data* and is available from the same publisher.

Until the development of this book, trying to piece together all of the necessary elements and processes that make up a remotely operated observatory was daunting. The authors and contributors have provided, in this single publication, a wealth of information gained from years of experience that will save you considerable money and countless hours in trying to develop such an observatory.

In these pages, you will be guided step-by-step through developing your observing plan design basis (OPDP), which will drive the selection of equipment to build the observatory. You'll also learn the discipline of "defense-in-depth and diversity" to maximize up-time and mitigate the possibility of a catastrophic failure while you are operating the observatory, possibly from a great distance, through your computer.

This book will make you take a hard look at the associated costs of making observations from a remote observatory. You will be able to weigh the pros and cons of building your own observatory versus buying observing time from a commercial facility. After careful consideration, you may find that your observing plan can be optimized by blending data collection from your own private observatory and a commercial observatory.

Amateur astronomers who understand how to properly collect and reduce science data can open up chances for themselves to access large telescopes and interact in the professional astronomy community. It should also be pointed out that there even are opportunities (albeit rare) for amateurs to be granted time on professional observatories. For example, in 1992, observing submissions from six amateur astronomers were selected by the Space Telescope Science Institute (STScI) that were carried out on the Hubble Space Telescope. The last section of this book presents an array of examples of work done by amateurs, professionals, and educators using remotely controlled observatories to achieve their observing goals. There is abundant inspiration there for you.

If you follow the methods and processes laid out in this book and choose to build your own remotely operated observatory or become a regular user of one of the commercial networks, you will not only join an elite group of advanced astronomers who make regular submissions to science but also become a member of an ancient fraternity. Your high-technology observatory will contain a "high-powered telescope" no matter how large it is, and from the comfort of home, you can actively contribute to the work that started in pre-history to help uncover the secrets of the cosmos.

Explore Scientific, LLC   Scott W. Roberts
Springdale, AR, USA
June 2015

Foreword II—
A Historical
Perspective

by Russell M. Genet, Ph.D.

In the past three and a half decades, since I first became involved with remote observatories, the use of remote, unmanned telescopes at fully automated observatories has advanced from a very rare approach for making astronomical observations to an increasingly dominant mode for observation among both professional and amateur astronomers. I am very pleased to see this timely book being published on the topic.

I highly recommend this book to readers because it not only covers the knowledge needed to become an informed user of existing remote observatories but also describes what you need to know to develop your own remote observatory. This book draws on more than two decades of remote observatory operation and networking by coauthor Richard Williams as he developed the Sierra Stars Observatory Network (SSON) into the world-class network it is today. This book is the ideal follow-on to coauthor Jerry Hubbell's book *Scientific Astrophotography* (Springer 2012).

As both a research astronomer and educator, I have a keen interest in involving my students in astronomical research. My Astronomy Research Seminar, offered through Cuesta College, provides both undergraduate and high school student teams an opportunity to plan, conduct, write up, and present a modest-scale, original research project within the constraints of a single semester. Our hybrid in-person/online spring 2015 seminar featured teams from ten schools: six in California, two in Hawaii, and one each in Arizona and Pennsylvania. Local volunteer assistant instructors, primarily high school science teachers, frequently met with the students in person, while I met with them online for instruction and student-team presentations.

It was vital that the teams made their observations early in the semester so they would have sufficient time for analysis and, more especially, for writing and rewriting their papers, which were submitted for publication by the end of the semester. Remote access observatories were the key to obtaining timely, high-quality, hassle-free observations for many of my student teams.

If, like my students, you would like to try your hand at scientific research or would just like to take photographs of the many fascinating objects in the night sky, an excellent and very affordable way to get started is to use remote observatories. This book, with its many practical examples, can be your guide.

Early Remote Observatories

I was asked by the authors of this book to provide a personal, historic introduction to remote observatories. For those interested in reading additional historic details, two classic books on the topic are still available on Amazon at low cost and are still helpful reads: *Microcomputer Control of Telescopes* (Trueblood and Genet, 1985) and *Robotic Observatories* (Genet and Hayes, 1989).

Before recounting my own involvement with robotic telescopes and observatories starting in the late 1970s—the early days of the Fairborn Observatory—it is appropriate to mention two earlier efforts in the 1960s that were, in essence, a space race between the National Science Foundation (NSF) and National Aeronautics and Space Administration (NASA). Which agency, by establishing the feasibility of robotic telescopes on the ground, would lead the way to robotic telescopes in space?

NSF purchased a 1.5-m Boller and Chivens® telescope that it located at Kitt Peak National Observatory in southern Arizona. The telescope was controlled by a mainframe computer located 40 miles away in Tucson. Automation was achieved, but reportedly for just one night. A mainframe computer was simply not a viable approach for controlling a remote observatory. The telescope was soon converted to manual operation, serving for decades as a workhorse near-infrared telescope. Recently, in a slightly ironic twist, it has been converted back to automatic operation, this time with a nearby microcomputer controlling the telescope. It might be noted that Sterling Colgate also developed a remote telescope to search for super-novae. The telescope, located on a dark-site mountaintop in New Mexico, was connected by a microwave link to a mainframe computer on the campus of New Mexico Institute of Mining and Technology, located in Socorro. It encountered the same sort of difficulties as the NSF telescope.

Art Code and his colleagues at the University of Wisconsin helped pioneer the path toward space telescopes with a modest 8-in. telescope located in a totally robotic observatory controlled from a nearby minicomputer, a PDP 8. While they were able to maintain fully automatic operation for several nights in a row without human intervention, the telescope was eventually shut down, although its space counterparts paved the way toward larger space telescopes.

With the arrival of microcomputers in the late 1970s and early 1980s, many robotic telescope developmental efforts were launched. Most were short-lived, but a few, such as the Fairborn Observatory telescopes and the Carlsbad Meridian Telescope, have continued operation to the present. Telescopes at the Fairborn Observatory began robotic operation in 1983 and remote access operation in 1987. The Carlsbad Telescope began robotic operation in 1984 and remote access operation in 1997.

The Founding of the Fairborn Observatory

In 1978, while attending graduate school on a full scholarship at the Air Force Institute of Technology (located at Wright-Patterson Air Force Base near Dayton, OH), I decided, on the side, to conduct some scientific research from my backyard that was both publishable and affordable. I quickly homed in on astronomical research and looked at every paper for the past 5 years that had been published in the *Astronomical Journal*. Not being a theoretician, I searched for papers about projects for which I felt I could make similar observations from my backyard (I lived a few miles from the small town of Fairborn, OH) with a telescope and instrumentation I could build myself. Somewhat surprisingly, there were 30 papers that fell into this category. Of these, 28 were photoelectric photometry observations of variable stars, primarily made by various observers using one of two 16-in. telescopes then at Kitt Peak National Observatory in Arizona.

Without hesitation, I ordered a set of 10-in. Cassegrain mirrors from Coulter Optical, two large-diameter worm gear sets from Tom Mathis (I was his first customer), and other items for the telescope. I constructed my photoelectric photometer from hobby plywood. It featured a diaphragm wheel, a filter wheel, a back-viewing microscope with retractable mirror, illumination of the back of the diaphragm that could be switched off when not in use, and an RCA® 1P21 photomultiplier. Rounding out the instrumentation was a high-voltage DC power supply that I built myself, a strip chart recorder, and a Hallicrafters shortwave receiver. Knowing that data reduction would be tedious (I am somewhat dyslexic when it comes to arithmetic), I purchased a Radio Shack® TRS-80 in early 1979.

Concrete was poured for the Fairborn Observatory's pier right after the spring thaw in 1979, and the first papers were sent off for publication that summer. Most all of the early observations were of RS CVn binary stars as part of a program coordinated by the late Douglas Hall at Vanderbilt Observatory. These binaries had large, dark, star spots on one hemisphere that produced light curves that changed over time as the large spots changed their locations.

Wishing to meet Doug Hall and other photometrists in person, I organized a June 1980 mini-conference that was held at the Dayton Museum of Natural History in Ohio. Doug stayed with me for a couple of days after the conference, and we

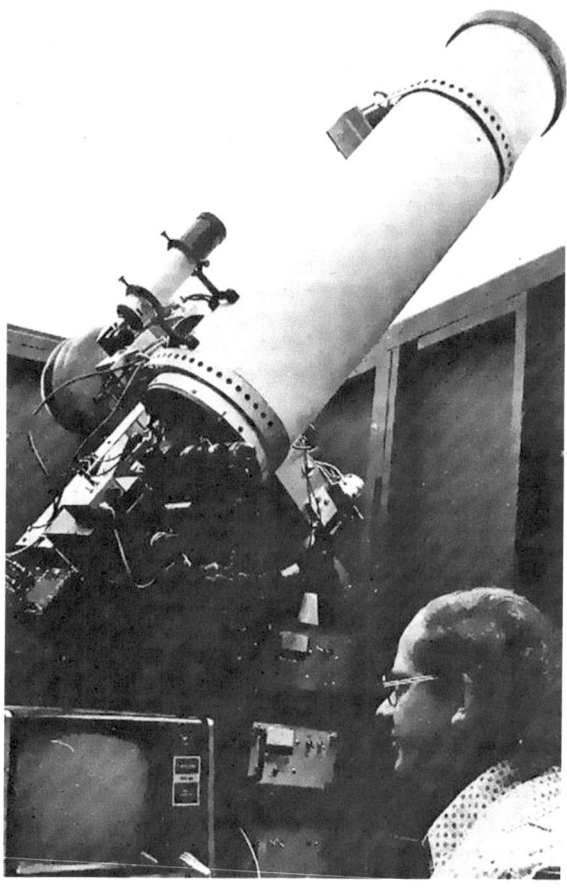

Fig. F1 Telescope, photometer, and Radio Shack® TRS-80 microcomputer at the Fairborn Observatory in 1979

launched the International Amateur-Professional Photoelectric Photometry (IAPPP) organization and its quarterly publication, the *IAPPP Communications*.

One good meeting deserved another, so I initiated a West Coast meeting in California that we called IAPPP West, and although I was unable to attend in person, the meeting took place and has been held annually for some 33 years, appropriately changing its name from the somewhat cumbersome IAPPP West to the much more informative Society for Astronomical Sciences. Back East, our next two IAPPP meetings were concerned with the use of microcomputers and resulted in two books, *Microcomputers in Astronomy I* and *Microcomputers in Astronomy II*, occasionally still available on Amazon. Doug Hall and I collaborated on a book, *Photoelectric Photometry of Variable Stars: A Guide for Smaller Observatories*. We were only slightly ahead of Arne Henden and Ronald Kaitchuck's masterful book, *Photoelectric Photometry*. Both books are still in print, although my book with Doug Hall was issued as a second edition.

Measuring strip chart traces of variable star observations was time-consuming and tedious, so I wired up an analog-to-digital converter to directly record the observations. Because the TRS-80 was not up to reliable operation in an observatory environment, I left it in my study and accessed it by a remote keypad and remote monitor. Soon a stepper motor was added to change the filter wheel. A good friend of mine, Johnathan Titus, author of the historic 1975 *Popular Electronics* article on how to build a microcomputer, asked me to write a book in 1982 entitled *Real-Time Control with the TRS-80*. As far as we know, it was the first published book on real-time control with microcomputers.

Early Automation

Making fully manual variable star photometry observations is repetitive, tedious, and boring. It consists of centering the same variable star, comparison, and check stars (and their sky backgrounds), over and over again in various filters. Strip chart recorder on, strip chart recorder off. Note the time. The observations went on and on, hour after hour. Partially computerizing the process by having a microcomputer directly record the data, change the filter wheel, and prompt me what to do next made the process easier and less error prone, but actually increased the boredom. Partial automation naturally increased the desire for full automation. Let the microcomputer do the complete job so we humans could sleep at night!

A photometrist friend of mine, Jeff Hopkins, mentioned that a friend of his, Louis Boyd, had similar thoughts of full automation of photoelectric photometry and kindly arranged for us to meet at Jeff's house during one of my trips to Arizona. Lou and I immediately hit it off, and we decided to join forces as the Fairborn Observatory East (my observatory in Ohio) and Fairborn Observatory West (Lou's observatory in Phoenix).

Lou's telescope was a very clever, home-brew contraption that used large aluminum disks (from an early mainframe computer disk drive) and bicycle chains for moving his telescope, and a single-board computer with a Motorola® 6809 processor for control. The computer had 64 K of RAM, but after loading the OS9 operating system and Basic09 language, there was only 18 K of RAM left for our control program. I went to Arizona for the first night of automatic operation of Lou's "Phoenix-10" telescope on October 13, 1983. We watched the telescope do its thing for a couple of hours and then went to bed. That first night, the Phoenix 10 made some 600 photometric measurements—finding, centering, and measuring stellar magnitudes in all three Johnson UBV color bands.

Some 6 months later, our second robotic telescope was in operation at my Fairborn Observatory East. The Fairborn-10's mount was donated by Frank Melsheimer at DFM Engineering®. It was the prototype of his line of mounts for smaller DFM telescopes. The 10-in. Schmidt-Cassegrain optical tube assembly was

Fig. F2 The Fairborn-10 automatic telescope. It began operation in the spring of 1984

donated by John Diebold, President of Meade®, while the photometer, an Optec® SSP-3, was donated by Jerry Persha. Lou Boyd kindly wire-wrapped the control board.

In the days before CCD cameras, our early robotic telescopes used their permanently mounted aperture photometers to locate a star (our mounts were not very precise in their pointing), center the star in a diaphragm, and, finally, measure the stars' brightness in various color bands. The stars were located by way of a square spiral search.

Once a star was located, it was centered by making four offset measurements and from these determining which direction the center must be. Successive iterations brought the star sufficiently close to the center of the diaphragm. Once centered, the first of a symmetrical sequence of some 33 separate observations was made that, over the course of about 11 min, yielded differential photometric magnitudes in three color bands.

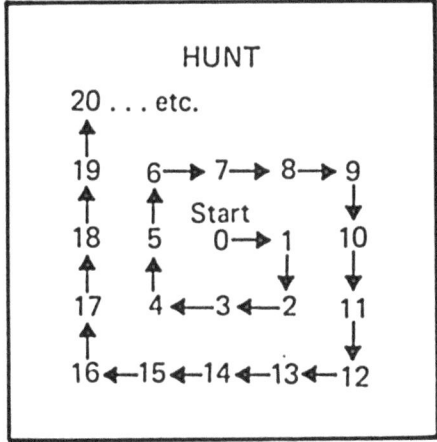

Fig. F3 The square spiral search routine used to locate stars

I attended the winter meeting of the American Astronomical Society (AAS) in Tucson in 1984. Dressed in shorts and a T-shirt, I called my wife back in Ohio every day. She was contending with −20 °F weather and freezing pipes. How about moving to Arizona, I suggested? Ohio, I had already concluded, was not a good location for robotic telescopes, and the government laboratory at Wright-Patterson Air Force Base where I was a research supervisor had a division in Mesa, AZ.

Immediately after the Tucson AAS meeting, Sallie Baliunas, an astronomer at the Harvard-Smithsonian Center for Astrophysics (CfA), took Lou Boyd, Doug Hall, and me on a tour of the telescopes on Mount Hopkins, south of Tucson. As we drove by a large roll-off-roof building, Lou asked about it and was told by Sallie that it was currently unused but was occupied by a laser ranger and Baker-Nunn camera that had been used in the past for satellite tracking. Lou insisted that we stop, and, although the building was locked, he paced out its dimensions.

Within months, I visited Dave Latham, the Director of the Smithsonian Observatory at CfA. We agreed that Lou and I could use the building on Mount Hopkins, along with utilities and use of Smithsonian Observatory vehicles, in exchange for devoting my Fairborn-10 telescope to making photometric measurements of solar-type stars for Sallie Baliuns to complement her spectroscopic observations on the 60-in. telescope on Mount Wilson. Together, these observations could establish "spot cycles" on other stars similar to our own solar cycle.

Not long after I moved to Arizona, I received a phone call from Dave Latham informing us that the Secretary of the Smithsonian Institution had approved a 10-year agreement for the Automatic Photoelectric Telescope Service, a joint

Fairborn Observatory–Smithsonian Institution operation with Lou and me supplying the telescopes and work, and the Smithsonian providing the building, utilities, and vehicles for navigating the long dirt road. The very next day, Lou and I drove to Mount Hopkins and bolted my Fairborn-10 telescope to the floor. Thus began the APT Service.

For more than a year, Lou and I spent many of our weekends and vacations on Mount Hopkins, automating the observatory itself. AC power on the mountain was not very reliable, so Lou, who was chief engineer for the telephone company, scrounged up an entire wall of batteries to provide us with the power we needed to close the massive roof when power failed. Weather sensors were devised, and Lou coded an observatory control computer to take charge of the observatory.

Finally, after much work, we arrived one weekend at the observatory, threw the switch to automatic, and let the observatory run on its own while keeping a careful eye on its operation. There were problems, of course, so these were fixed, and we tried it all over again the next weekend. This went on for several months until at last we had a number of weekends of fully automatic operation in a row without any problems. Somewhat nervously, we finally left the switch on "automatic," left the observatory running on its own, and drove back to the Phoenix area, a 4-h drive from Mount Hopkins.

It wasn't too many days before we received a call from one of the daytime crew members of the Multiple Mirror Telescope (MMT). "As I was driving by the APT Service, I noticed that your roof was rolling off, then rolling back on, then rolling off ..." We asked a friend on the mountain to stop automatic operation. Once the software fix was made, the observatory was back in automatic operation again.

It was somewhat nerve wracking always wondering how our automatic telescopes were doing. We would often call down to our friends, the nighttime operators of the Smithsonian's 60-in. telescope just down the ridge from our APT Service, and ask them to walk up to our observatory and check on our automatic telescopes, now three in number. These frequent calls were somewhat disruptive, so Lou programmed our Observatory Control Computer to send us a "morning report" over the Internet that summarized the operation of each telescope as well as the observatory itself for the previous night.

These helpful morning reports usually informed us whether or not another weekend trip to Mount Hopkins was required, although there were occasional conflicting reports. Once, for several mornings in a row, the morning report illogically reported that the skies were clear but it was raining. A call to our friends on the mountain established that it had been clear all the time. We drove down, and an inspection of the rain detector revealed that a bird had used our sensor as a toilet facility.

Our observatory eventually settled down into routine, reliable operation, really cranking out the observations. My Fairborn-10 telescope observed solar-type stars for Sallie Baliunas. The newly acquired Vanderbilt-16, a DFM Engineering® telescope funded by NSF, was devoted to observations of RS CVn spotted binaries for Doug Hall. Lou Boyd's Phoenix-10 was devoted to the APT Service's "Rent-a-Star" program. For $2.00, we made UBV photometric observations (33 separate

Fig. F4 The original three robotic telescopes at the APT Service on Mt. Hopkins. *Left* to *right*: the Fairborn-10, the Phoenix-10, and the Vanderbilt-16

measurements spread over about 11 min) of a variable star, comparison, and check star (and associated backgrounds). Our Automatic Telescope Instruction Set (ATIS) was smart enough to decide what should be observed next, given the desired frequency of observation, number of past observations, location of the Moon, etc. If poor weather shut the observatory down for a few hours, once started up again, ATIS would carry on with its selections, always picking what was most appropriate to observe at any given time.

At this point, our biggest problem was implementing the changes in the observing program requested by the many astronomers using our telescopes and then distributing the observational results back to them. This problem was particularly severe for the Phoenix-10 Rent-a-Star telescope, which had many subscribers. We decided that each telescope should have a Principal Astronomer (PA). The PA would consolidate the observational requests from multiple users, make updates to the observing program, check the observations for quality, distribute the results to other astronomers, and, for the Phoenix-10 telescope, collect observational fees. Michael Seeds, at Franklin and Marshall College, was the first PA, boldly taking on the difficult Phoenix-10. Sallie Baliunas, at the CfA, was the PA for the Fairborn-10, while Greg Henry, at Tennessee State University (TSU), was the PA for the Vanderbilt-16, taking care of it for Doug Hall. PAs worked out well indeed, shifting the workload to those who were in the best position to handle it.

One final problem remained to be solved, and that was timeliness, both in terms of the inspection of the observational results to determine whether their quality remained high and in terms of making changes to the basic observing programs. Observational results were stored, untouched, on 3.5-in. floppy disks for 3 months. At the end of each quarter, a disk from each telescope, containing a quarter's worth of observations, was mailed to each PA, who reduced the data, checked its quality, and distributed the reduced results to other astronomers. The PAs also sent occasional observing program changes to us, which we loaded on their telescope.

We were always worried that sometime during the quarter, a subtle equipment fault would occur that could not be detected until the data were reduced. To speed up this process, we instituted remote access in 1987, perhaps the first such access to telescopes. Observational program change requests made by PAs over the Internet during the day would be honored that night. The night's observational results were automatically sent via the Internet every morning to each of the PAs so they could reduce the observations and check their quality.

Fairborn Observatory Expansion and Maturation

The IAPPP (East) annual summer conferences were moved to Arizona in 1985 and became mid-winter conferences that were held for many years at the Lazy K-Bar Guest Ranch near Tucson. These conferences were attended by an eclectic mix of professional and amateur astronomers. (Many of the amateurs were professional engineers or computer programmers.) The Lazy K-Bar conferences featured research and development talks, sunbathing around the pool, horseback riding in the desert, and annual pilgrimages to the growing collection of robotic telescopes at the Fairborn Observatory on nearby Mount Hopkins.

Many of the talks at these annual conferences were published in a series of books. A number of these books are still available at low cost on Amazon and provide insights into the early development of robotic telescopes and remotely accessed observatories. A film crew documented the development of robotic telescopes during a 3-day visit to the Fairborn Observatory that was aired on PBS as "The Perfect Stargazer."

The telescopes on Mount Hopkins were shut down every summer because of severe lightning during the monsoon season. However, the cool summer mountaintop environment was ideal for meetings, and a series of workshops were devoted to advancing robotic telescope and remote observatory technologies. During one of these workshops I made a scale model of a 0.8-m telescope, specifically intended for automated photometry. Soon a number of these telescopes, designed in detail by

Fig. F5 Four 0.8-m automatic telescopes on Mt. Hopkins. The three original telescopes are in the background

Lou Boyd, were manufactured by Rettig Machine in Redlands, CA. Four of these telescopes were installed at the Fairborn Observatory on Mount Hopkins.

A regular speaker at the Lazy K-Bar IAPPP conferences was Wes Lockwood, an astronomer from Lowell Observatory. Wes made highly precise photometric observations on a 21-in. telescope that were a cut above robotic telescope in terms of photometric precision. I figured that robotic telescopes should be able to make observations that were at least as precise as mere humans, so I organized and hosted two workshops devoted to figuring out how we could make much more precise automatic observations. The workshop attendees included Wes Lockwood, Bill Borucki from NASA Ames (who was working on making photometric observations precise enough to detect exoplanet transits), Andy Young from San Diego State University (who was an expert on sources of error in photomultiplier aperture photometers), Lou Boyd a seasoned instrument designer, myself, and a few others. Based on what Lou learned in these workshops, he designed a fully automated photometer that produced observations significantly more precise than the best manual observations.

Downloading each night's observations every morning made it possible for the PAs to institute daily quality checks on the observations. Greg Henry at TSU developed a semi-automatic quality check he used every morning to look for potential problems in the several telescopes for which he was the PA. At the first hint of a problem, Greg would contact Lou, who would immediately investigate and take corrective actions as appropriate.

In 1991, Bill Borucki and I proposed using a network of robotic telescopes equipped with high-precision photometers to detect exoplanet transits. There was little interest in this at the time—after all, we were not even sure there were any exoplanets, let alone transiting exoplanets—so the idea didn't go anywhere. However, Bill was already thinking about placing a high-precision automatic photometric telescope in space to search for exoplanets.

Alex Filippenko, at the University of California, Berkeley, had a keen interest in investigating the expansion of the universe by observing Type 1-a supernovae. The problem was finding these rare, transient supernovae. Alex figured that a robotic telescope would be a good way to find them. By taking images of many galaxies every night and comparing them with previous images, he hoped to see whether a bright supernova had shown up. Alex came to several of the Lazy K-Bar conferences, and arrangements were made to build one of the 0.8-m telescopes for him as the Katzman Automatic Imaging Telescope (KAIT). It was installed at Lick Observatory in California and held, for a number of years (late 1990s and early 2000s), the world's record for the number of supernovae discoveries. The knowledge gleaned from these Type 1-a supernovae observations was helpful in establishing the baseline for the interpretation of Type 1-a supernovae observations that led to the discovery of the runaway universe.

In the late 1990s, Geoff Marcy asked Greg Henry to use one of the 0.8-m automatic photoelectric telescopes at the Fairborn Observatory to look for potential exoplanet transits. While none had yet been observed, Geoff had a good feel for which of the exoplanets, discovered via small changes in the radial velocities of the parent stars, might produce a photometrically detectable transit. Greg Henry observed the first exoplanet transit in November 1999. David Charbonneau made simulation transit observations of the same exoplanet, and Dave's paper was published immediately following Greg's in the *Astrophysical Journal*.

At a number of the Lazy K-Bar IAPPP annual conferences, there were talks and discussions on the automation of spectroscopy. Eventually, Mike Busby at TSU obtained funding for a 2-m automatic spectroscopic telescope, which was located at the Fairborn Observatory in its new location in southern Arizona, just 5 mi north of the Mexican border. Currently, this telescope is equipped with very high-precision (just a few meters/second), fully automatic, fiber-fed spectrometers from TSU (Matthew Muttersburgh) and the University of Florida (Juan Gee).

Fig. F6 Two-meter automatic spectroscopic telescope at the Fairborn Observatory

The Fairborn and Other Remote Observatories Today

In the early 1990s, I retired from my job as a laboratory supervisor, after having worked for 33 years for the federal government and also having completed my term as the 52nd President of the Astronomical Society of the Pacific. I retired from the Fairborn Observatory, and Lou Boyd became its sole Director. I had greatly enjoyed my 15-year exploration of observational astronomy and robotic observatories, but felt the call to return to and continue my research, lectures, and courses on cosmic evolution, the grand synthesis of physical, biological, and cultural evolution. For those who might be interested, my book on this topic, available from Amazon, is *Humanity: The Chimpanzees Who Would Be Ants.*

Lou Boyd at the Fairborn Observatory and Greg Henry at TSU carried on just fine without me. The observatory continued to grow. Today, more than two decades after I retired, Lou continues to manage the Fairborn Observatory as a one-person

Fig. F7 The Fairborn Observatory now features some 14 automatic telescopes. The 2-m spectroscopic telescope can be seen in the background

operation, easily making it the most cost-efficient observatory on the planet. A dozen telescopes make observations every clear night. Greg continues as the planet's most experienced PA of robotic telescopes. He has authored or coauthored hundreds of papers with his many collaborators in the many research areas where automatic photometric and spectroscopic telescopes excel.

The approach of having multiple telescopes under one roof was picked up, early on, by Mark Trueblood at the Winer Observatory southeast of Tucson. His remote access Winer Observatory has been serving many astronomers for decades. A recent example of a collection of a dozen telescopes under one roof is demonstrated by iTelescope in Australia.

Individual telescopes were also automated, many inspired by my 1985 book with Mark Trueblood, *Microcomputer Control of Telescopes.* Sophisticated telescope control systems, such as the one made by Sidereal Technology®, became available off the shelf. High-performance telescope mounts also became off-the-shelf items, such as the Paramounts made by Software Bisque®, and complete robotic telescopes became available, such as the Corrected Dall-Kirkham telescopes made by PlaneWave Instruments®. Many of the ever-popular Schmidt-Cassegrain telescopes made by Celestron® and Meade® were also automated.

Controlling the telescopes, cameras, weather sensors, roofs, and all the other things that made up a complete remote observatory remained a difficult challenge until Bob Denny developed the Astronomer's Control Panel (ACP) and established the Astronomy Common Object Model (ASCOM) as a standard interface for astronomical devices managed by a personal computer. Starting in the early 2000s, Bob programed and then continuously refined his telescope and observatory control software for both remote real-time and fully automatic operation.

One of our early dreams was a network of robotic telescopes spread around the globe. A number of these networks have now been established, including the Sierra Stars Observatory Network (SSON) established by Rich Williams, one of the authors of this book. There are now many other networks, such as the AAVSOnet and the iTelescope.

The premier global network is the Los Cumbres Observatory Global Telescope (LCOGT) network, established by Wayne Rosing after he retired from Google® as its Vice President for Engineering. The LCOGT network has two 2-m telescopes (one on Haleakala on Maui, and one in Australia), ten 1-m telescopes (in Chile, South Africa, Texas, and elsewhere), and a dozen 0.5-m telescopes, collocated with the 1-m telescopes. This matched set of telescopes and instruments is controlled automatically from LCOGT headquarters in Goleta, CA (not far from Santa Barbara).

A few years ago, I became curious to learn about the present state of remote observatories. Because I spent every winter in Hawaii, the dry side of the Big Island seemed like a good place for a conference. I had not gotten very far in my organizational efforts when Christian Veillet, then the Director of the Canada-France-Hawaii Telescope (CFHT), volunteered to handle the conference's organization. The result was the *Telescopes from Afar* conference. Christian kindly invited me to give the first talk on the early history of robotic telescopes and observatories. Christian knew everyone, and some 150 astronomers from all over the world attended and reported on their uses of robotic telescopes and development of new capabilities.

Remote observatories have a bright future, opening up astronomy to a new and much larger generation of professional, amateur, and student observers. Machines and humans can and do work well together. I hope you enjoy reading this book as much as I have and will take advantage of the developments over the past several decades by the many pioneers of remote observatories.

California Polytechnic State University   Russell M. Genet, Ph.D.
San Luis Obispo, CA, USA

Russ is also Distinguished Visiting Professor of Astronomy, Concordia University Irvine; Adjunct Professor of Astronomy, Cuesta College; Adjunct Professor of Space Science, University of North Dakota; and Editor, *Journal of Double Star Observations*.

Preface

Welcome to *Remote Observatories for Amateur Astronomers: Using High-Powered Telescopes from Home*. The idea for this book came about through discussions with Maury Solomon, Nora Rawn, and John Watson of Springer Books. Our objective was to create a follow-on book to *Scientific Astrophotography: How Amateurs Can Generate and Use Professional Imaging Data* that was not only timely but also provided significant new material to enable the amateur astronomer to expand his or her horizons and delve into the latest techniques and equipment available today.

It immediately became obvious to me that to provide the highest quality material, this new book would require the involvement of an expert in remote observatories; I had just the fellow in mind. Rich Williams is well-known throughout the astronomy community as a pioneer in designing, building, and operating remote observatories and operates the Sierra Stars Observatory Network (SSON). This network of state-of-the-art observatories was built in 2007 and has been in operation ever since.

I also realized that to provide you, the reader, with the quality reading experience you expect, an excellent editing job was paramount. In this regard, I asked the editor of my previous book, Linda Billard, to join the team to provide her expert skills and knowledge to meld the material that Rich and I would be delivering. This ensures the book speaks with "one voice," avoiding the distractions that different authors' writing styles can cause for readers. Linda has a wealth of experience in this regard, and I am very grateful for her work on this book.

This book, which is intended for those astronomers who are interested in learning all the details of designing, building, and operating a remote observatory, provides the resources to get you quickly up to speed on what is involved in building your own remotely operated observatory or working with an existing commercial remotely operated observatory service. This book provides a wealth of detail not only on the systems, subsystems, and components (SSC) that make up the remote observatory but also how to integrate those SSCs and operate the observatory in a professional manner.

In *Scientific Astrophotography*, I introduced the concept of an Astronomical Imaging System (AIS) to describe the design and function of the SSCs that provide data based on a specific observing program. In this new book, we expand on this concept and discuss how it may apply to your specific remote observatory design and to the various observing programs you want to run as the observatory director. This approach also offers insight for those astronomers who need to match their specific observing programs to the equipment provided by the remote observatory services available today to astronomers all over the world.

Having access to your own remote observatory is a significant milestone in the life of an astronomer whether you own it or rent it. It will make your life significantly easier in several ways, including not having to set up or tear down your AIS, being able to work in a comfortable environment, and efficiently taking advantage of those observing opportunities that quickly changing weather may have denied you in the past. Productivity goes up as frustration goes down, and your life as an astronomer becomes golden.

Using a commercial observatory service allows you to focus on your data rather than equipment issues and operations if that is your desire, and again boosts your productivity. These are advantages particularly valuable for astronomers involved in scientific observing programs. Providing accurate and timely observations to those scientific groups that accept them are key to developing a reputation as a professional-caliber astronomer, whether you are paid or not for your high-quality work.

It is my expectation that this book will be the definitive "go-to" book on remote observatories long into the future and will help astronomers—whether they are amateurs or professionals—who are looking to move to the next level in their astronomy "career."

Lake of the Woods Observatory MPC I24   Jerry Hubbell
Locust Grove, VA, USA
June 2015

Acknowledgments

We would like to thank several amateur and professional astronomers for their contributions to Part III of this book, the case study section. Part III provides you, the reader, with a strong sense of what is possible when using a remote observatory and presents a wide variety of observing programs on which to model your own. Specifically, we would like to thank Adam Block, Roger Dymock, David Galbraith, Kevin Healy, Carl Hergenrother, Rob Matson, Patrick Miller, Robert Mutel, Kevin Paxson, David Pulley, Derek Smith, Americo (Eric) Watkins, and George Faillace for their valuable contributions to this book.

We would also like to thank Scott Roberts and Russ Genet for their generous offer to provide the forewords for this book. Their points of view are based on their vast experience in the industry and years of working with amateur and professional astronomers all over the world. We would also like to thank Rachel Konopa for her excellent work on the book cover.

Finally, we would like to thank Springer Books for this opportunity to create a book that we believe is needed at this time of great change in the industry to help amateurs keep abreast of the many new ways they can enjoy their hobby. We thank our Springer editor Nora Rawn for her support and understanding while assembling the manuscript; it is greatly appreciated.

Contents

Part II   Using Remote Observatory Facilities

List of Figures

List of Tables

About the Authors

Gerald R. Hubbell is currently the Director of Electrical Engineering for Explore Scientific, LLC, and an Assistant Coordinator of Topographical Studies, Lunar Section, for the Association of Lunar and Planetary Observers (ALPO), and former president of the Rappahannock Astronomy Club (raclub.org). He is the author of the book *Scientific Astrophotography: How Amateurs Can Generate and Use Professional Imaging Data* (Springer 2012). He has more than 30 years of experience in the nuclear utility industry as an expert in nuclear instrumentation and nuclear process controls and protection. He has been an amateur astronomer since his teenage years and has been active for more than 5 years in modern astrophotography.

Richard J. Williams is the founder and CEO of the Sierra Stars Observatory Network (SSON), a global network of professional remote observatories available to everyone. Starting in the mid-1990s, he was a pioneer in the development of robotic telescope hardware and software. He was a co-founder of Torus Technologies (now Optical Mechanics, Inc.), which designs and manufactures robotic telescope systems and custom optical-mechanical devices for government and industry. He owns and operates the 24-in. (0.6-m) Sierra Stars Observatory telescope located at his ranch in California, which was the first telescope to go online for SSON in 2007.

Linda M. Billard is a freelance technical writer/editor with 30 years of experience. Her clients span FORTUNE 500 companies to small businesses. Her focus is developing electronic and hardcopy products sensitive to the technical level of the reader. Products include documentation, marketing collateral, books, and newsletters; proposals; and software requirements analysis and design. Her interest in astronomy ramped up about 6 years ago when she became active in the Rappahannock Astronomy Club (RAClub). She is the editor/contributor for RAClub's well-regarded online newsletter, *StarGazer*, and its online presence at www.raclub.org, and the technical editor of *Scientific Astrophotography: How Amateurs Can Generate and Use Professional Imaging Data* (Springer 2012).

Part I

What Is a Remotely Controlled Observatory?

Chapter 1

Introduction to Remote Observatories for the Amateur Astronomer

Today, the Internet makes vast amounts of information publically available to you in the form of databases, news feeds and blogs, informative websites, social media, and so on. Amateur and professional astronomers use the Internet to communicate and collaborate in ways that were unimaginable only a few decades ago.

After the Internet opened up to private and commercial entities in the mid-1990s, professional and amateur astronomers began using it to communicate and disseminate information and data. They created websites to inform the public about their programs and projects; formed online forums to communicate among themselves; and communicated directly via email, instant messaging, social media, and voice over IP (VoIP). What once were groups of astronomers working in isolated remote-mountain and backyard observatory sites became an interconnected global community. This disruptive technology changed how people do astronomy in revolutionary ways. One notable consequence is the renaissance of amateur astronomers doing valuable astronomy projects in collaboration with the professional astronomy community. These citizen scientists do serious work, ranging from taking images using their own telescope equipment; to measuring and analyzing image data, looking for patterns and hidden information in the vast accumulating data collected by space-based and ground-based observatories; to writing articles and blogs about recent developments, and more.

About the same time, the Internet became public, amateur astronomers began using the first commercial charge-coupled device (CCD) cameras for esthetic and scientific imaging. This is arguably the most revolutionary observational technology advance in astronomy in the twentieth century. In the 1980s, using small backyard telescopes with CCD cameras, amateur astronomers could take images that rivaled the results from much larger professional observatory telescopes taking images with photographic films.

© Springer International Publishing Switzerland 2015
G.R. Hubbell et al., *Remote Observatories for Amateur Astronomers*, The Patrick Moore Practical Astronomy Series, DOI 10.1007/978-3-319-21906-6_1

Advances in computers, electronics, software, and materials over the past few decades are now incorporated into modern telescopes. As a result, telescopes and even entire observatories can operate automatically throughout an entire observing session, taking hundreds of images without human intervention. This automated process eliminates the tedious tasks required to operate an observatory manually and optimizes the amount of image data collected in an observing run. Now, available high-speed Internet connections enable people to control observatories remotely in a direct or queue-based manner from the comfort of their office or home.

Together, all these converging technologies enable many more people to participate in doing astronomy. A growing number of professional-quality remote observatories now serve colleges, institutions, and individuals. Some enterprising people have made professional observatory telescopes commercially available to anyone willing to pay a small fee to operate a telescope remotely either through direct control or by scheduling images to be taken and delivered to them.

Remote Observing—Pushing Down the Technology

Today, you, the amateur astronomer, have access to a broad range of information and technology, as well as the help and service of a large community of experts and professionals willing to spend time helping you quickly get up to speed on doing professional-level work if you so desire. In addition, hundreds if not thousands of other likeminded astronomers (amateur and professional) enjoy the beauty of the night sky and have perfected the art of acquiring and processing awe-inspiring images using not only backyard telescope systems, but also professional data from the Hubble Space Telescope (HST) and other space- and ground-based professional observatories. The Internet is largely responsible for our ability to share this vast amount of data and observations from astronomers all over the world.

Over the past 15 years, and particularly over the past decade, the technology used by the professionals has steadily decreased in cost, moving it into the affordable realm for the amateur astronomer. Computers and computing technology have enabled you to continuously increase the performance and accuracy of the data generated by your backyard telescope, or astronomical imaging system (AIS), and allowed new astronomers everywhere to quickly learn the techniques and acquire the skills to contribute substantially to this avocation that we all love and pursue.

The data are acquired mainly using CCD cameras, which, over the past decade, have increased in performance and in the sheer amount of data you can collect with them. The sensitivity of today's CCD cameras exceeds the capabilities of the equipment that the professionals were using even a decade ago, and the data that amateurs are acquiring today rival what the largest observatories in the world were gathering even in the 1980s. Coupled with the professional-level techniques that have been developed over the past 20 years, you can provide data to those professional and amateur organizations that appreciate the work you are doing and they will archive and make your observations available for use by astronomers worldwide.

Another major contributor to the increased value of observations made by amateurs is the availability of extremely large databases and catalogs of astronomical objects that make it much easier to identify those objects on the images you acquire and report on the constantly changing condition of those objects. In particular, in the area of minor planet research, accurately measuring the changing position (astrometry) and brightness (photometry) over time of asteroids contributes greatly to our overall understanding of the evolution of the solar system and its planetary members. Using the UCAC4 (USNO CCD Astronomical Catalog 4) enables you to very precisely match the objects in your image to the catalog and quickly identify any stray objects in your images that are then identified using the Minor Planet Center database of minor planets. If you are very lucky, you may identify a new object never seen before. In that case, you may be credited with a new discovery and given the chance to name the object.

The astronomical community has many contributors who volunteer their time developing state-of-the-art software applications to process the raw data we acquire with our AISs and distill the information down to something that is most useful to the community at-large. These applications use the large professional catalogs and also professional techniques that have been developed and perfected over the past 20 years to provide the consistent level of data quality necessary for use at the professional level.

The most recent developments have further enhanced the use of your AIS. Typically, the small amateur telescope has been a portable system that is set up at the beginning of the observing session and then taken down at the end of the session. Usually, setup takes about 30–60 min and then subsequent polar alignment, sky alignment, and data acquisition of the calibration frames take additional time at the beginning of the evening. Then, after several hours of use, the system is dismantled and stored in the appropriate protective containers, which again takes up to 60 min. In addition, you will be running your AIS directly, out in the elements, for several hours. So, for a portable system, add 2–2.5 h to a nominal 4- to 8-h observing session.

Over time, this somewhat time-consuming and tedious setup and take-down process tends to add to your reluctance to set up your AIS unless the weather is predicted to be really good for observing. (The seeing is good or better, and the transparency is very good, although for science imaging, this is not necessarily a requirement.) The end result is that you do less observing rather than more. The solution, of course, is to set up your system in a permanent or semi-permanent state protected by an enclosure that can be opened on demand. Having your system ready to go and configured for your observing program for use in 10 min or less goes a long way toward increasing the number of observing sessions. Because the technology is available today, adding the components to enable you to remotely operate your AIS from within a warm room or even in your office makes it that much more enjoyable. A remotely operated AIS, whether it is local or located up to several hundred or thousands of miles away, is available to you today at a reasonable cost.

The ultimate version of this type of observing operation is available through an observatory service owned by a third party. This arrangement allows you to rent time on several different AISs. You pick the system and schedule your observations to be performed by that system. There are also systems available that allow you to directly control the AIS remotely with live feedback using video monitoring and live data feeds. Either way, you can operate someone else's system for a fee without having to invest the time to design, build, and operate your own equipment. This may be the best choice if you want to learn the professional techniques of gathering and processing your own data without the large investment in time and money it takes to build your own system. Another advantage is that if your observing program works well with your own AIS but you need some other data that only a much larger telescope can provide, then the use of an observatory service can be a lifesaver.

As you can see, there are a large number of options available to the amateur astronomer today and myriad choices available to meet your requirements and the needs of your observing program. Over the past couple of decades, the technology has increasingly been pushed down to lower-cost systems, and today is available to you whether you are a very experienced astronomer or are just starting out learning about the sky.

This book shows you how remotely controlled observatories are designed, how the components that make up the typical remotely located observatories are specified and integrated into your AIS, and finally how they are operated and made available to astronomers all over the world.

Do-It-Yourself Remote-Controlled Observatories

As mentioned in the section "Remote Observing—Pushing Down the Technology," the state of the art in personal observatories has advanced to the point where you can operate your remotely controlled AIS from the comfort of your office or anywhere in your home you choose to be. Many of you are interested not just in using a remotely controlled observatory, but in buying and integrating your own components into subsystems and fully operational AISs. This is very doable with the diverse and readily available commercial off-the-shelf components and systems available today for you to integrate into your own remotely controlled observatory.

There are basically two types of do-it-yourself observatories—remotely located and locally located. Typically, if you are used to setting up and taking down your AIS for each observing session, your next step is to set up a locally located observatory to save you the time and effort of setting up and taking down your portable AIS. You observe with this observatory just as before—by sitting next to your AIS and operating it hands-on (Fig. 1.1).

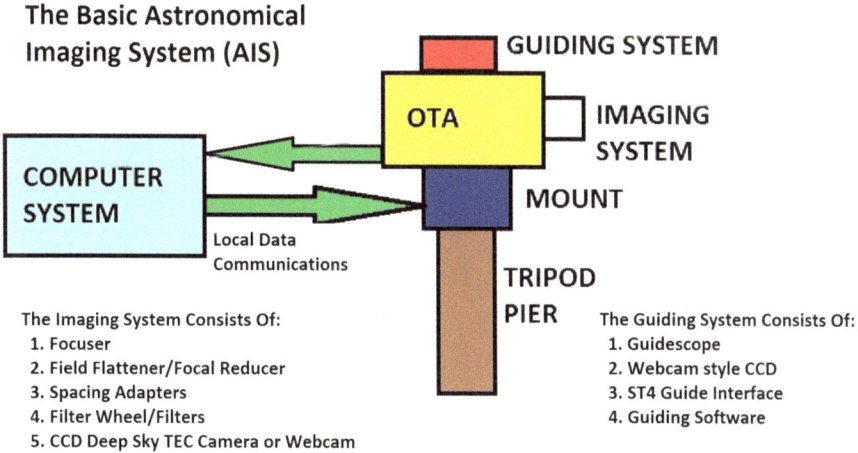

The Basic Astronomical Imaging System (AIS)

GUIDING SYSTEM

OTA

IMAGING SYSTEM

COMPUTER SYSTEM

MOUNT

Local Data Communications

TRIPOD PIER

The Imaging System Consists Of:
1. Focuser
2. Field Flattener/Focal Reducer
3. Spacing Adapters
4. Filter Wheel/Filters
5. CCD Deep Sky TEC Camera or Webcam

The Guiding System Consists Of:
1. Guidescope
2. Webcam style CCD
3. ST4 Guide Interface
4. Guiding Software

Fig. 1.1 Schematic of a portable, hands-on AIS

Of course, this still presents several issues, principally the fact that you still have to sit outside in the elements. One advantage is that if any of your AIS equipment has problems or fails in any way, then you are on-site (so to speak) and you can correct it immediately. While this is a good thing, if you have a few years of experience with your equipment, you have worked out any bugs in your hardware and software configurations, and probably have the confidence to leave your equipment alone for a period of time while it is acquiring data. If you have reached this point in your operations, then you may want to consider an alternative.

Once you have your permanent observatory set up and operating reliably to the point where you do not have to babysit your equipment, it is not that much more effort to add the subsystems necessary to remotely operate your equipment in the comfort of your home office. This process primarily consists of physically separating your computer/control equipment from your field instruments and locating the computer/control equipment in your office (Fig. 1.2).

The two separated systems are connected to each other via a high-speed communications link that gives you real-time control of your remotely operated AIS instruments. There are other enhancements you may want to make even if you are within a few yards of your AIS instruments, including upgrades to your power supply system and addition of some monitoring instruments, including a web camera (webcam) so that you can watch your AIS in operation from your home office.

Part I of this book details all the things you need to know about the systems, subsystems, and components (SSC) used in building a remotely controlled observatory system, and also those extra items needed to make sure the completed AIS is reliable and available at a moment's notice to observe those transient objects that require a timely observation. Part I also includes valuable information to help you get the most from your system.

Remotely Operated
Astronomical Imaging System (AIS)

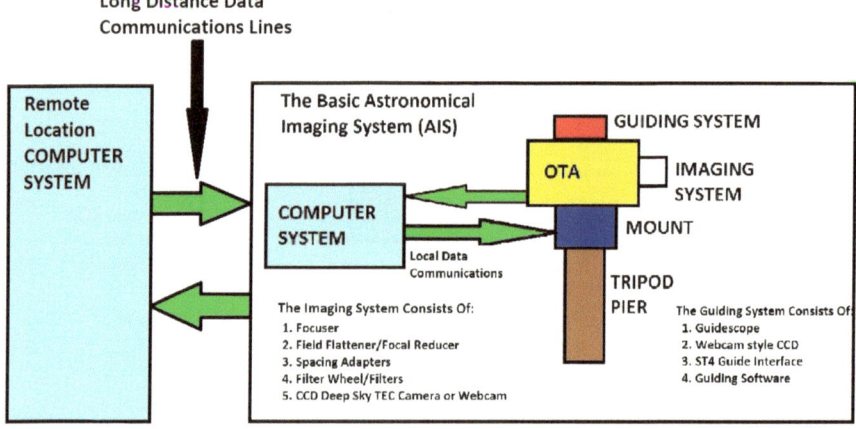

Fig. 1.2 Schematic of a remotely operated AIS

What You Can Do Using a Remote Observatory

Once you have your remotely controlled observatory up and running, you can use it to do all the things you did previously while sitting next to your AIS. If you are into the scientific study of astronomical objects, then you should understand the specific objects (asteroids, planets, deep sky nebulae) that you want to observe, and design your equipment configuration and imaging train to match your object (Fig. 1.3).

By having access to a remote observatory configured and set up for quick startup and shutdown, you can concentrate on your observing program instead of worrying about your equipment configuration every time you have an observing session. This frees you to focus on the quality of your observations and allows you to collect much more data on more objects, contributing to your depth of knowledge of the astronomical objects of your choice.

Also, your remote observatory allows you to observe more often when the weather permits because you can start up your remote observatory in a very short period of time, typically within 10 min, and start executing your observing program. By collecting more data, more often, on a given astronomical object, you can contribute more to the greater astronomical community and accomplish your goals at a higher quality than ever before.

This book delves deeply into all the different projects you can do using a remote observatory. Part II provides information on how to use and operate a remote observatory system that is available over the Internet. There are basically two types of

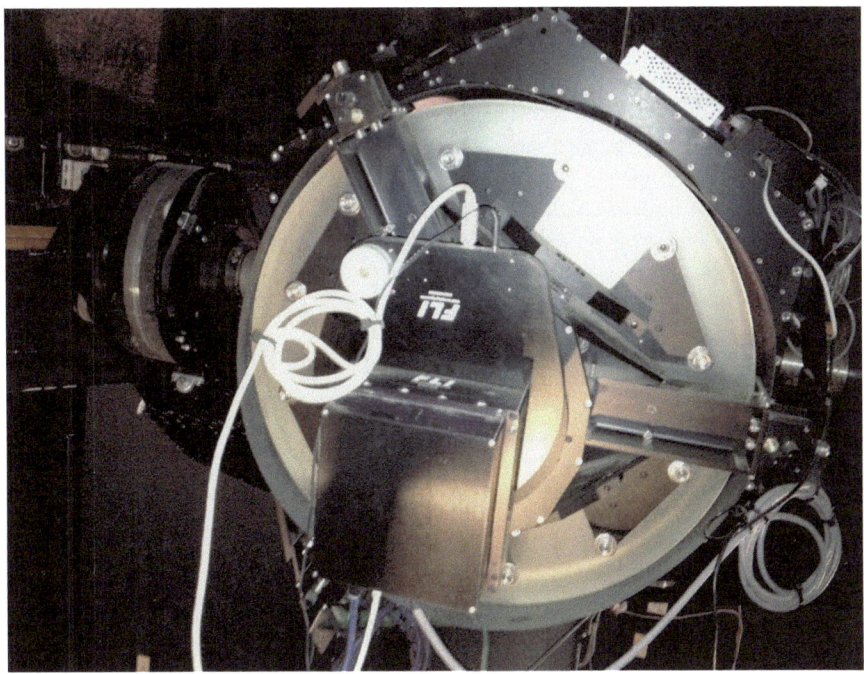

Fig. 1.3 A remotely operated observatory at work

commercially available remote observatories—those that you control directly in real time and those to which you submit an observing request that is scheduled for you. The latter systems automatically perform the data gathering and then typically email you when your data are available to download to your local computer system.

Part II also includes the information necessary to help you decide whether you should build your own observatory or take advantage of the AIS systems available to you over the Internet. Weighing your costs in time and money is an important consideration in this decision, and depending on your long-term goals, you may be surprised at what you end up deciding.

Part III of the book contains case studies describing projects that have been done using remote observatories by amateurs and professionals alike. These provide all the details of the science behind the observations, the equipment used, and the analysis performed to discover new things about these astronomical objects. These case studies include the information you need to help you decide what objects interest you and what is possible using a remote observatory in your observing program. If you plan to design and build your own remote observatory, you will find everything you need to help specify the systems, subsystems, and components necessary to create your AIS to duplicate the work discussed in the case studies.

Whether you build your own, or use a commercial remote observatory, you will be embarking on a noble quest to gather unique and fresh data on various astronomical objects; sharing that data is one of the best things you can do for the astronomical community. Remember that we, amateurs and professionals alike, are all in this together, and because we are a small community, we need to support each other the best that we can.

Overall, this book provides you with all you need to get started in designing, building, or operating a remote observatory to further your personal goals in astronomy.

Further Reading

Arditti D (2007) Setting-up a small observatory. Springer
Berry R, Burnell J (2005) The handbook of astronomical image processing. Willmann-Bell
Buchheim R (2007) The sky is your laboratory. Springer
Chromey FR (2010) To measure the sky. Cambridge University Press
Covington MA (1999) Astrophotography for the amateur. Cambridge University Press
Dragesco J (1995) High resolution astrophotography. Cambridge University Press
Dymock R (2010) Asteroids and dwarf planets and how to observe them. Springer
Hubbell GR (2012) Scientific astrophotography: how amateurs can generate and use professional imaging data, Springer
Smith GH, Ceragioli R, Berry R (2012) Telescopes, eyepieces and astrographs. Willmann-Bell

Websites

Cherry Mountain Observatory, www.cherrymountainobservatory.com
iTelescope, www.itelescope.net
LightBuckets, www.lightbuckets.com
New Mexico Skies, www.nmskies.com
Sierra Remote Observatories, www.sierra-remote.com
Sierra Stars Observatory Network (SSON), www.sierrastars.com
Slooh, main.slooh.com
University of Iowa Robotic Observatory (Rigel), astro.physics.uiowa.edu/rigel
University of Arizona Mt. Lemmon SkyCenter, Skycenter.arizona.edu
Warrumbungle Observatory, www.tenbyobservatory.com
Winer Observatory, www.winer.org
ADM Accessories, http://admaccessories.com/
Apogee Imaging Systems, http://www.ccd.com/
Astro Tech, https://www.astronomytechnologies.com/
Astro Hutech/BORG, http://www.sciencecenter.net/hutech/
Astro-Physics, Inc., http://www.astro-physics.com/
Atik Cameras, http://www.atik-usa.com/
Celestron, http://www.celestron.com/
Daystar Filters, http://www.daystarfilters.com/
Denkmeier Optical, Inc., http://www.deepskybinoviewer.com/
Explora-Dome (Poly Tank), http://www.exploradome.us/

Explore Scientific LLC, http://www.explorescientific.com/
Finger Lakes Instrumentation LLC, http://www.fli-cam.com/
Hotech Corp, http://www.hotechusa.com/
Howie Glatter's Laser Collimators, http://www.collimator.com/
Innovations Foresight LLC, http://www.innovationsforesight.com/
IOptron, http://www.ioptron.com/
Lunt Solar Systems LLC, http://www.luntsolarsystems.com/
Meade Instruments, http://www.meade.com/
Moonlite Telescope Accessories, http://www.focuser.com/
Peterson Engineering Corp, http://www.petersonengineering.com/sky/index.htm
Planewave Instruments, http://www.planewaveinstruments.com/
QHYCCD (Astro Factors/Deep Space Products), http://www.astrofactors.com
QSI, http://qsimaging.com/
Questar, http://www.questar-corp.com/
Santa Barbara Instrument Group, http://www.sbig.com/
Shelyak Instruments, http://pagesperso-orange.fr/shelyak/en/index.html
Sky Watcher, http://www.skywatcher.com/
Software Bisque, http://www.bisque.com/
Starlight Instruments LLC, http://www.starlightinstruments.com/
Starlight Xpress Ltd., http://www.sxccd.com/
Stellarvue, http://www.stellarvue.com/
Tele Vue Optics, http://www.televue.com/
Texas Nautical/Takahashi America, http://www.takahashiamerica.com/
Vernonscope, http://www.vernonscope.com/
Vixen Optics, http://www.vixenoptics.com/
William Optics, http://www.williamoptics.com/

Chapter 2

Remote Observatories: Land-Based Space Probes

What Is a Remote Observatory?

When discussing the use of a remote observatory, it is important to understand exactly what that term means. A remote observatory contains an Astronomical Imaging System (AIS) made up of two distinct subsystems: the AIS instrumentation and the AIS control system/human-machine interface (HMI). Typically there are two different configurations of the subsystems, depending on how close the two are located to each other. For observatories where the two subsystems are within 100 ft of each other—a locally located remote observatory—the AIS instrumentation contains the power system, dome controller, astrograph, charge-coupled device (CCD) camera(s), filter wheel, focuser, mount controller, and any miscellaneous instruments, such as perhaps an electronic finder and a webcam to keep an eye on the AIS (Fig. 2.1).

The separate AIS control subsystem contains the master computer that runs the software, which interfaces to all the instrumentation via a hard communications cable of some kind, either a powered universal serial bus (USB) cable system or perhaps a network cable system. The AIS control subsystem is located in a warm room such as a home office setting or other suitable location in the home or office. The AIS control subsystem computer has all the commercial off-the-shelf (COTS) and custom software installed and running to operate the observatory.

Observatories that are designed to be used at long distances via the Internet— known as remotely located observatories—have a different configuration. Typically, the AIS at the remote location contains those items listed previously for the AIS

© Springer International Publishing Switzerland 2015   13
G.R. Hubbell et al., *Remote Observatories for Amateur Astronomers*, The Patrick Moore Practical Astronomy Series, DOI 10.1007/978-3-319-21906-6_2

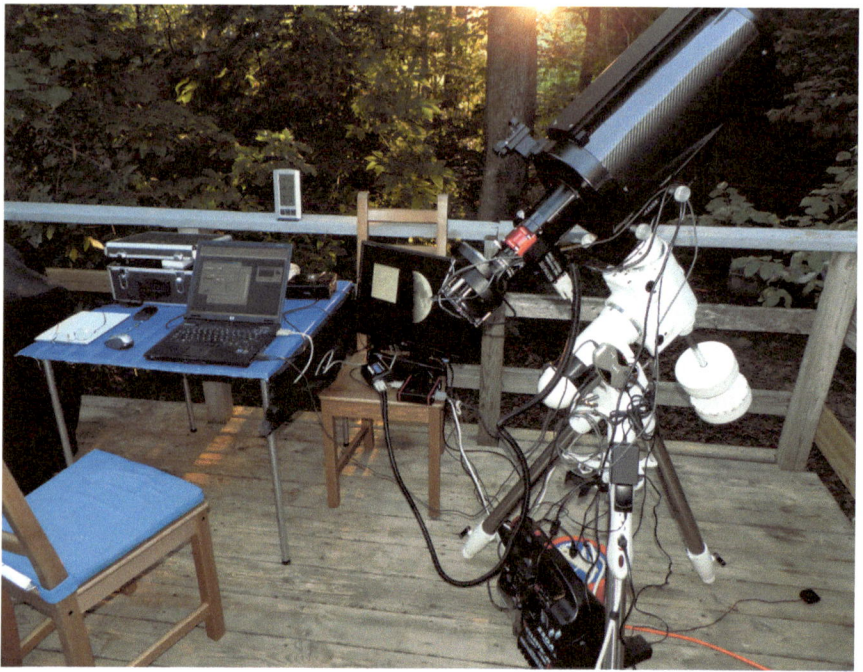

Fig. 2.1 Locally located remote observatory system

instrumentation subsystem plus a master computer system that contains the COTS and custom software used to operate the observatory (Fig. 2.2).

The AIS instrumentation subsystem also contains components that provide some level of redundancy for reliability purposes; chief among these is a redundant power supply and communications components. Typically, additional software is installed to provide a generic HMI accessible using an Internet browser. This includes a web server and a data storage subsystem with remote access capability. All of the above is located at the remote observatory. The main access point is provided through the Internet, which allows operation from anywhere in the world. This AIS is further defined by the type of operation that is permitted, specifically real time or scheduled (Fig. 2.3).

Real-time operation is just as it sounds; you control the observatory directly as you desire. You send commands through the HMI, and you observe the results of your commands in real time. This is typically how a locally located AIS is controlled also. A scheduled operation consists of an HMI that allows you to select the target, define the timeframe for the observation, select the instrumentation to be used, and then submit these requirements to a queue for automatic processing by the system.

Fig. 2.2 Remotely located remote observatory system

Real-time versus Scheduled Observatory Operations

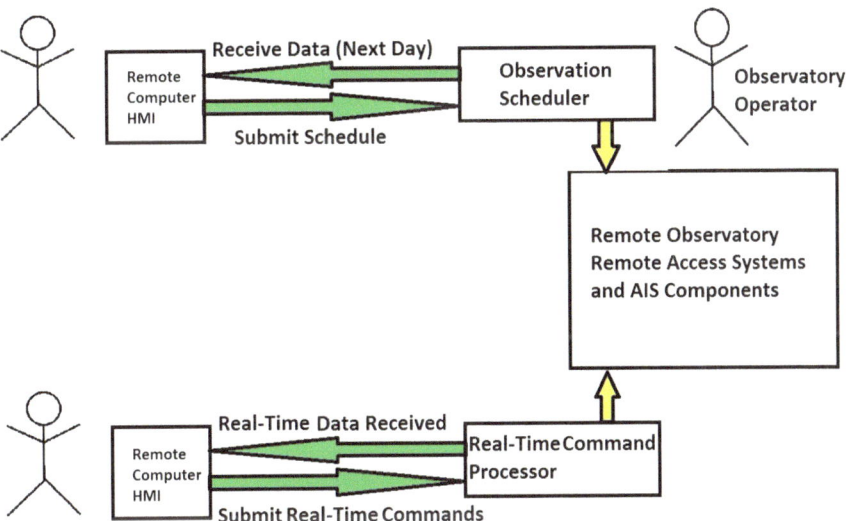

Fig. 2.3 Real-time versus scheduled observatory operation

Why Use a Remote Observatory?

Why should you consider the use of a remote observatory? That's a good question. Most amateur astronomers start out by purchasing their own equipment and through the years are able to purchase new items to add to their system. Once you have a few years under your belt, you may be interested in doing science with your AIS. Your AIS should provide you with an excellent system with which to learn all the procedures and acquire the skills and knowledge to start doing science observing. Soon you will realize that you may need to acquire some new instruments that may be outside your budget. This is when using the services of a commercial observatory can come to the rescue. There are several good reasons for using a remotely controlled observatory, whether it is one you built or a commercial service. These include:

1. Both allow quick startup and shutdown times, making it easier to observe when the weather is good.
2. The commercial observatory provides instruments that you do not have or cannot afford to acquire.
3. Commercial providers offer professional-quality telescopes and instruments located at observatory sites with dark skies and excellent seeing conditions.
4. Both the amateur and commercial observatory provide a long-term consistent configuration that collects excellent, high-quality, consistent data.
5. Both allow you to do observations in the comfort of your home or office.
6. The commercial observatory provides professional quality astrographs in sizes that allow you not only to perform follow-up work but also discover new objects.
7. Using a commercial service allows you to spread the cost out over several months or even years as you get your data over an extended period of time because there are no upfront costs.
8. Building and owning your own locally located observatory engenders a pride of ownership that using a commercial service does not. There is nothing quite like owning your own equipment, regardless of the cost, time, and effort it takes to operate and maintain it.

These are the primary reasons that most amateurs decide to use a remotely controlled observatory. Designing and building your own allows you to have the best of both worlds, the pride of owning your own observatory along with the advantages of a commercial service. Although this can be a challenge, it is well worth the effort to understand what it takes to build your own remotely controlled observatory.

Before you can use a commercial remote observatory service, you need to develop your observing program to identify the astronomical objects that you are going to study. Part III of this book provides you with case studies of a variety of observing programs that astronomers are executing using these services. Once you have identified the objects you want to observe, then you need to determine the instrumentation required to observe them. The following sections and chapters will help you do this.

Most commercial remote observatory services have systems optimized for imaging faint objects. They have high quantum efficiency CCD cameras designed to take long exposures of faint objects. Because the minimum shutter speed on these types of CCD cameras is about 0.1 s, bright objects, such as the Moon and major planets, are overexposed even at the fastest shutter speed. To get high-resolution lunar and planetary images, your best option is a video camera system designed to take a large number of frames per second (fps) (typically 30–120 fps) and to store them in a video file format, either an .avi or other uncompressed format. Special software is available to process these video files into beautiful high-resolution lunar images that rival those that professionals were obtaining just a couple of decades ago (Fig. 2.4).

The major limitation for commercial services in supporting lunar and planetary imaging is the very large video files acquired for each observing run. Typically, each video file consists of 500–2000 frames, which can be up to 1 gigabyte (GB) in size. Amateurs often acquire 20–30 GB of video in one evening using their own equipment. Transferring large video files with many gigabytes of data from commercial remote observatory facilities would be impractical for most services at today's typical Internet bandwidth rates. Larger commercial telescopes, ranging in size from 20 to 32 in. (0.5 to 0.8 m) with long focal lengths, can provide high-resolution, large image scale lunar and planetary images if the cameras have shutter speeds fast enough not to overexpose the objects.

What Does It Take to Build a Remote Observatory?

If you have the skills and knowledge to operate a sophisticated AIS that you are comfortable setting up and taking down for each observing session, then you should not have any problems turning your system into a remotely controlled observatory. If you have already mastered the selection, integration, and operation of the components and systems you use for your imaging, you likely only need to add a few components to complete your remotely controlled AIS.

As mentioned previously, two main ingredients are reliable power and reliable communications. Because the instruments in a remotely controlled AIS are usually separated from the control system, you need two power systems, one for each subsystem. The best power is typically AC main power provided by a power utility. This is the preferred primary source of power for both the AIS instruments and the control system. As a backup power supply for your AIS instruments, you can install an alternate source of power such as a battery coupled to an AC inverter (Fig. 2.5).

Most equipment available to you today (cameras, focusers, mounts, etc.) can be powered directly from a 12 VDC battery. You can power computer and networking equipment directly from a battery system if you use some kind of voltage converter. Some laptops require 18 VDC, and other computer equipment requires 5 VDC. It may be easier (although not nearly as efficient) to use an AC inverter from the battery to provide 120 VAC to the various power adapters for each of the computers or network components. Parts of your power systems are exposed to the local environment while your observatory is operating. Therefore, you need to

Mare Humorum
Gassendi to Vitello

Jerry I lubbell
Lake of the Woods Observatory
Locust Grove, VA

01/12/2014 2326 UT
FL 4864mm f/32
Explore Scientific 152 ED APO CF
Televue 4x Powermate
Point Grey Flea3 GigE CCD
Red FIlter
Processed with Registax6, MaxIm DL,
and Photoshop 6.0

Fig. 2.4 High-resolution lunar image

consider how you will keep your battery systems warm enough to last through an observing session in the coldest conditions and keep all power system components dry and protected from any rain or moisture to prevent corrosion and failure of the power connectors.

Remote Observatory
Power System Layout

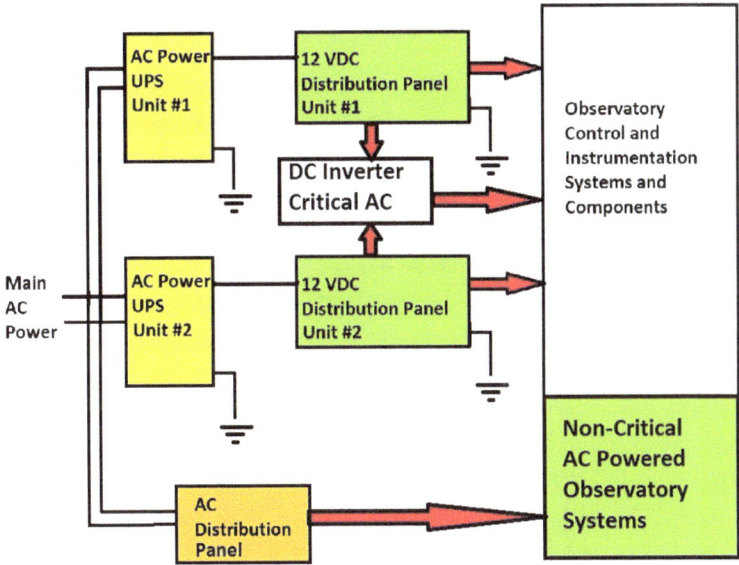

Fig. 2.5 Remotely controlled observatory power system layout

You should use a hard-wired communications system, such as a high-speed USB cable or Ethernet network connector. If you use Ethernet, then the instruments you buy must have an Ethernet network interface available. USB interfaces are more common, although more and more instruments are being delivered with network capability. For a robust, reliable connection, avoid any wireless devices. Although convenient, they are not as dependable as hard-wired connections.

Pay special attention to the integrity of your connectors and other connections in your AIS instruments and control system (Fig. 2.6). For permanent power connections, hard-wired cabling using terminal boards and screw terminations should be standard. These connections provide the highest reliability and ensure continued dependability of the AIS. Cable connectors for USB, serial, Ethernet, and other interfaces should have screw-type locks and also some form of strain relief in case a cable is pulled in some way.

You must route the cables around any moveable parts of your AIS, including the mount and focuser, which typically are the only moveable parts of your system. Make sure you test the full range of motion to ensure your cables and connections are free to move without binding. Make sure you protect all your connections from moisture. Connector boots are an effective way to protect your connections from dirt and moisture.

Communications
Connector

Power Connectors

Fig. 2.6 Reliable power and communications connection types

Finally, you should consider providing lightning protection for any communications and power lines that run outside between your locally located AIS instrumentation and the control system components. These cables are susceptible to lightning strikes, which can propagate high currents in the affected power and/or communications lines. This, in turn, can destroy any equipment terminated on those lines. A close lightning strike can overwhelm the best power insulation devices in a system. To ensure your system is completely isolated from a major electrical surge during a lightning storm, develop a power cable switching system that completely disconnects the power from every electrical device associated with your AIS, including all such equipment both inside and outside. An Ethernet connection or USB Internet connection to your computer could conduct an electrical surge to your system that could destroy sensitive electronics.

A Land-Based Space Telescope (Reliability, Functionality, Monitoring, and Safety)

One unique way of looking at the design and construction of a remotely located observatory, where the goal is the reliable and robust performance of your AIS, is to treat it like a land-based space telescope that it is both expensive and painful to gain physical access to for maintenance and repairs. Only in this way can you begin to understand the scope of the needed robust design features that will keep your AIS operating day in and day out.

Consider the design requirements of a satellite in orbit around the Earth. The broad areas to consider include power systems, energy management, safety, environmental protections, consumable resources, mission requirements, monitoring instruments, communications, and data storage. The systems for a remotely located observatory have most of the same systems because they are operated as though they were unreachable, just as a satellite in Earth orbit is. Instead of mission requirements, you have an observing program design basis that dictates the requirements for instrumentation. The main consumables for your land-based system are power and storage space on your computer disk drives. Fortunately, you have no systems that use liquid or gas consumables as satellites do for maintaining station and attitude.

Power Systems

As stated previously, the power supply for a remotely controlled observatory must be reliable and dependable. The most cost-effective way is to connect your hardwired main AC source to a battery backup system, commonly called a UPS or uninterruptible power supply. You can buy UPS units with varying power capacities. The backup power capacity you require depends on the amount of instrumentation and control system equipment you have, which dictates the total load you need to power. The typical configuration is to power all your instrumentation and control system from the 12 DC bus (Fig. 2.5).

This way you can easily measure your total load and usage. You can use a digital multimeter to measure the total load of your system when you mock up your configuration. Use care when making the power connection through your multimeter and make sure you have it on the correct setting before powering up your system. Enlist the help of an experienced person if you have any questions about how to do this measurement.

Let's say you have a total load requirement of 2 amps (A) at 120 VAC—this is equal to 240 VA (volt-amps). At a power factor of 0.9, this is equal to 216 Watts-AC (WAC). Using a conservative value of 85 % for efficiency (most systems are typically greater than 90 % installed) boosts the requirement from 216 WAC to 255 WAC. The per hour rating on batteries is standard, so if you use a 12 VDC battery system UPS, then you require 255 WAC/12 VDC, or 21 A per hour. A 750 amp-hour (Ah) deep-cycle marine battery at 80 % capacity would last approximately 28 h.

Because you have two locations to power (the AIS instruments are separated from the AIS control system), you need a UPS system for each location rated according to the load at that location. Store each UPS in a warm, dry location. The remote AIS instrumentation is typically outdoors and exposed to cold temperatures, so an insulated enclosure warmed with a single 60- or 100-W lightbulb inside will help to maintain the UPS at a workable temperature. Finally, there is one other component that you may consider installing—a remote power switch. It allows you to switch your main power off and on via a communications link, either Ethernet or USB. These switches allow you to power down and restart your whole system in case of an unforeseen lockup of the AIS.

Observatory Power Measurements

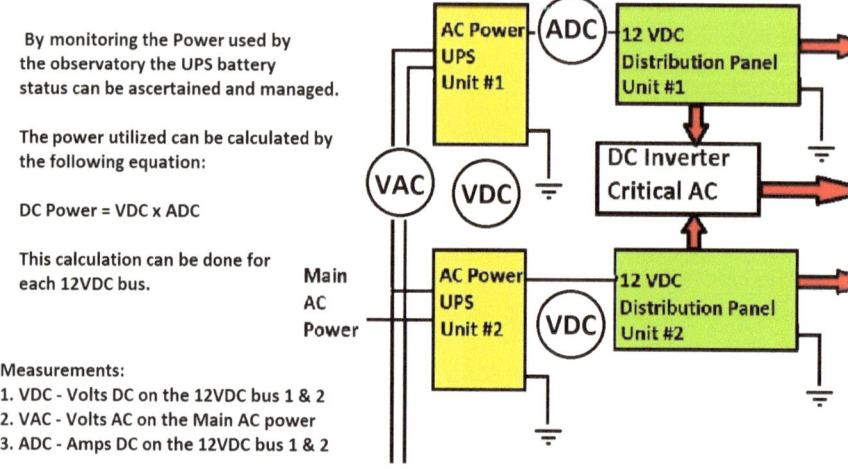

By monitoring the Power used by the observatory the UPS battery status can be ascertained and managed.

The power utilized can be calculated by the following equation:

DC Power = VDC x ADC

This calculation can be done for each 12VDC bus.

Measurements:
1. VDC - Volts DC on the 12VDC bus 1 & 2
2. VAC - Volts AC on the Main AC power
3. ADC - Amps DC on the 12VDC bus 1 & 2

Fig. 2.7 Energy management using basic monitoring instruments

Energy Management

It is important to monitor your power supply system and manage your energy usage. You can accomplish this by using a simple set of instruments to monitor your power supply voltage, current, and temperature (Fig. 2.7).

Because you have two locations to monitor, you have a minimum of eight instrument readings: two 120 VAC measurements, two battery temperature measurements, two 12 VDC measurements, and two 12 VDC total system amp measurements. Simply multiply your battery voltage by the 12 VDC total system amp measurement to get your amp-hour usage. Divide your battery's amp-hour rating by the amp-hour usage to get your estimated hours. To be more conservative, multiply this result by 0.8 to leave a margin for error.

By using these instruments and monitoring them during your observing session, you can identify any problems early on, which allows you to easily discontinue operations and reliably shut down your system to eliminate or at least minimize any damage to your equipment.

Safety Monitoring

Safety monitoring is really all about making sure that while operating your AIS, you do not cause any damage to people or property, specifically you or your remotely controlled observatory. The goal is to keep yourself and your equipment safe to operate another day.

Because you are not in the vicinity of your equipment when operating it, you need to have a way to keep monitoring the operation remotely to make sure you do not move your astrograph into a compromising position, or leave your AIS exposed to the elements when rain or snow is forecast. A simple solution is to use a webcam to view your equipment and use your control system to monitor the status of all your instrumentation to preclude any adverse movements or operations.

Environmental Protection

Environmental protection focuses on making sure that your remote observatory's equipment is operated within its required temperature and humidity range. As previously discussed, the power supply system has special requirements in this regard because it has a battery subsystem. That was discussed in the section "Power Systems." Your remotely controlled observatory building is likely either a dome or roll-off roof system. Both systems work well to protect your telescope and instruments from the outside weather.

By using additional monitoring instruments, you can guard against accidental exposure to rain, snow, and ice. An all-sky camera and a weather station greatly enhance your situational awareness regarding the local weather and atmospheric conditions. By using additional software and Internet resources, you can monitor the current and forecast weather conditions to support your observing sessions planning (Fig. 2.8).

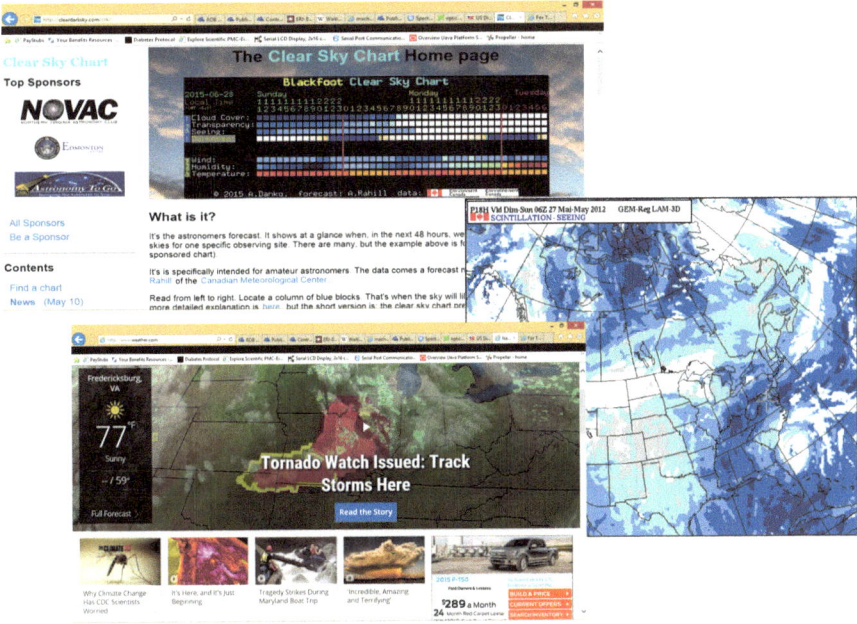

Fig. 2.8 Weather-monitoring services and software

Because remote observatories are open to the outside environment during observing runs, pests such as rodents and insects have easy access to exposed equipment. Rodents can damage wiring by chewing on it and also can damage telescopes by building nests inside them. Insects, such as wasps, can also build nests inside, which can be a pain (literally) to clear out without damaging your equipment. To mitigate any damage that these pests can cause, you must remain vigilant and inspect your facility regularly. These pests can cause intermittent issues that lead to permanent failures if not caught in a timely manner. Unfortunately, this is one environmental concern that you must deal with down here on the ground that satellite systems do not encounter. Diligence in monitoring for pests is a requirement for maintaining a robust, reliable, and dependable system.

Consumable Resources

For your purposes, the only consumables that you may have to manage are power usage and cost, and data storage system usage and cost. Fortunately, the price of electricity is fairly stable. In the United States, power is priced anywhere from about $0.12 to $0.20 per kilowatt-hour (KWh). In the example discussed in the section "Power Systems," the 255 WAC at a nominal cost of $0.16 per KWh would cost $0.49 per 12 h of operation. If you were lucky enough to be able to operate your system for 15 days during each month, then your monthly electricity usage would only amount to $7.50. In the scheme of things, considering the cost of your AIS instruments and control system, plus ongoing storage costs, this is next to nothing.

On the other hand, data storage needs can amount to quite a sum depending on the type of astronomical objects you want to observe and how much of the raw data you want to keep. There is a huge difference between observing lunar and planetary objects, and deep sky objects. As mentioned previously, photographing lunar and planetary objects requires the acquisition of high-speed video data to tease out the high-resolution features that you are after. It would not be unusual for you to acquire 10–30 GB of raw, uncompressed video data over just one observing session. At the nominal average rate of 20 GB per session a 2 TB (terabyte) disk drive would be full after just 100 observing sessions. If you were lucky enough to be able to observe 10 nights per month, the disk drive would last about 10 months. These drives currently cost about $150–200 (in 2015). If you want to provide for failures, then a redundant pair of drives would cost twice that, or up to $300 per 10 months, or $360 per year. Perhaps in the future, there will be other archival systems that are more cost effective, but currently, there is no other way to store that much raw data in a cost-effective manner. There are specific advantages to keeping all that raw video data, chief among them the ability to reprocess the video when newer, better processing software becomes available to tease out even more details from the raw data. It is your decision. You can save a substantial amount of money, and vastly extend the life of your storage system, by deleting your raw video data after processing them into the final images that you want to keep.

Deep sky imaging is much more reasonable in terms of required storage space. Typically, you will use exposures of 1–10 min, where 1-min exposures are the worst case in terms of data storage requirements because you will need to make more of them. For a deep sky camera that uses the Kodak® KAF-8300 CCD chip, each frame requires approximately 17 megabytes (MB) of storage space. If you acquire 50 frames per hour (i.e., slightly fewer than one per minute) for 4 h in each observing session, that amounts to 200 frames or 3.4 GB of data. At this rate, if you observe 15 nights per month, you won't use up your 2-TB disk drive for 39 months or a little over 3 years. If you spend $300 for a redundant disk pair, then the cost per year is approximately $92. Remember, this is the worst case. If you typically take 5-min exposures, then this cost is reduced by a factor of 5, with a resulting cost per year of only $18, and you won't fill up your disk for 95 months or a little over 16 years. As you can see, storage costs for deep sky work are much lower, and you can definitely keep all your raw image data without much concern about storage space.

Observing Program Design Basis

When you set out to design your remotely controlled observatory, one of the first things you want to do is decide what type of astronomical objects you want to observe and record. This starts with deciding what kind of science you want to do. Examples include astrometric measurements of asteroids, photometric measurements of variable stars, photometric measurements for light-curve measurements, or spectroscopic measurements of stars. Each one of these types of measurements requires a specific set of instruments that will optimize your results and provide the most bang for your invested dollar. The book *Scientific Astrophotography: How Amateurs Can Generate and Use Professional Imaging Data*, published by Springer Books, goes into great detail on this point.

Once you have decided on your science goals, and designed and identified your AIS components, then you can assemble your remotely controlled observatory. You should always strive to make sure the science objectives drive your equipment requirements—this ensures you get the best raw data you can obtain with the level of equipment you can afford. You can always decide to have multiple science goals, but make sure that the equipment you use to pursue a specific science goal is the best you can afford. Always change out the equipment on your AIS to support the best data acquisition you can.

AIS System Monitoring

Monitoring your system for proper operation is an ongoing task that you will become accustomed to during your observing session. As discussed in the section "Power Systems," monitoring your power system is of prime importance to make sure that your system does not shut down prematurely in the middle of your observing session. In addition, it is important to be able to see your AIS movements as you progress through your imaging. Watching for cable issues and other problems that result from the mount's movements will help ensure trouble-free operation. A simple and inexpensive way to maintain awareness is to use a webcam to watch these movements. Two webcams are even better and can help reveal problems that using only one may not allow you to see.

Installing an all-sky camera to monitor the sky's cloud cover is a good idea because it will help you to know when it is necessary to shut your system down before a quick-moving storm can affect your equipment.

Communications Systems

Your communications system is as important as your power supply system for reliable, uninterrupted operation of your remotely controlled observatory. Your communications system is the backbone of your whole AIS because it connects all your instrumentation to the control system's various software and hardware components. As discussed in the section "What Does It Take to Build a Remote Observatory?", the communications system brings together the two separate parts of your remotely controlled observatory—the AIS instruments and the control system.

To create a robust communications hardware system, you should consider building a redundant, parallel, configuration consisting of duplicate cable runs, routers, USB hubs, and any other hardware needed. You can set up this redundant communications link as a standby system that you can turn on when the primary system fails. Alternatively, you can use both systems as a communications load-sharing system where the AIS instruments and control system software components are distributed between the two communications systems. This second option allows for soft failures of individual system components, making the failures manageable, versus a hard failure of the complete communications system, resulting in a total loss of control of your remote equipment.

A remotely controlled power switch on your system, as discussed in the section "Power Systems," gives you the flexibility to power down and restart important components. This requires a totally separate communications line between the two locations because you may need to restart the redundant pair of communications systems along with the AIS instrumentation and control system components. This third communications cable gives you that capability, which may be needed at some point in the operation of your remotely controlled observatory.

Storage Systems

Depending on the type of observing you do, your needs for data storage can be very large. Settling on a standard 2-TB hard drive system installed in your control system computer gives you the space and necessary data transfer speed to record high-speed video for your high-resolution lunar and planetary imaging. If your observing program focuses on deep sky imaging of astronomical objects, then you have the option to use the existing smaller disk drive installed on your control computer and adding a large 2-TB network-attached storage (NAS) subsystem on which you archive your raw data. This option also allows you to have a disk subsystem that has a pair of redundant disk drives configured as a single mirrored disk. This guards against any one disk failure where your raw data are archived. Additional NAS disk subsystems can be added at any time in the future as needed.

If your control computer is a laptop, the internal battery acts as a backup power source in case the main AC power is interrupted. This ensures that the installed hard drive will not shut down prematurely. If you use a NAS or other external disk drive, then you need to provide for a redundant power source via a battery system. Plugging these external drive systems into a UPS ensures that the drive systems are not shut down prematurely.

Internet Connectivity

Internet connectivity to your control system is not a primary system in terms of the needed components to meet your observing program and can be considered a support system that may or may not be available. This is not the case if you are building a system that will be remotely located tens or hundreds of miles away from your control location. In that case, your only means of high-speed communications may be a secure and robust Internet connection. This will be discussed later.

Here the focus is having an Internet connection locally on your control system to access the World Wide Web and email for communicating with third parties and making notifications. Unless you specifically choose to do so, your observing program procedures should not rely on the Internet for any of your data processing. Of course, if you already have high-speed Internet at your control location, then by all means, use it but do not rely on it for your observatory operation.

As you are probably aware, your science goals drive the design of your observing program, which in turn drives the design of your AIS and the specific components to efficiently acquire the data to answer your science questions. After a few years of operating your portable AIS, you have more than likely decided what astronomical objects really interest you and those that do not.

There are probably two or three areas or questions that you would like to investigate and thus include in your Observing Program Design Basis (OPDB). The book *Scientific Astrophotography: How Amateurs Can Generate and Use*

Professional Imaging Data goes into great detail on how to create your OPDB and design your AIS to meet your observing goals.

If you expend the effort to create your own OPDB, then you will realize cost savings when you build your remotely controlled observatory. This process will ensure you have all the equipment you need, and nothing extra, to realize your science goals. Make sure you make the effort to sharing your data and results with those groups that encourage your contributions, such as the Minor Planet Center (MPC), the Association of Lunar and Planetary Observers (ALPO), and the American Association of Visual Star Observers (AAVSO).

Part III of this book details examples of the science projects you can do using your remotely controlled observatory. There are a wide range of projects you can do, and depending on your AIS instruments, you may be able to observe several different types of astronomical objects using the same equipment.

How the Remote Observatory Increases Productivity

By now you have realized that having a remotely controlled observatory will save you hours of setup and breakdown time for your AIS and also provide you a comfortable environment to work. It is well worth the effort, if you want to own your own observatory, to spend the hours to design and build your remotely controlled observatory. Having the flexibility to start an observing session at a moment's notice is a liberating feeling and is conducive to being able to observe more and make better use of your AIS equipment.

Once you have operated your observatory manually, you may want to add the necessary software and hardware to more fully automate your system. In doing so, you can start your AIS running and then get some well-deserved sleep. There is nothing like having your AIS operate automatically all night and wake up in the morning to find a wealth of raw data ready to be processed and analyzed.

At some point, you might find that your existing equipment or observing site is inadequate to fulfill all your observing goals. If you reach that point, you have the option of using one of the available commercial remote controlled observatory systems. Part II of this book describes these systems in detail.

Is Having My Own Remotely Controlled Observatory for Me?

Designing and building your own observatory is not an easy task. Add to that the requirement to have it remotely controlled, and it can be a great challenge. If you have experience integrating all the components to do imaging with your portable AIS, then you probably have the skills to convert your system to a remotely controlled one. Alternatively, you may want to continue to use your portable AIS as you do presently, and augment your work using a commercial remotely controlled observatory service. This is a good way to understand what you are missing with your portable observatory and may prompt you to go ahead and build your own. Either way, you will find remotely controlled systems both challenging and rewarding, providing hundreds of hours of satisfying work.

Further Reading

Arditti D (2007) Setting-up a small observatory. Springer
Berry R, Burnell J (2005) The handbook of astronomical image processing. Willmann-Bell
Buchheim R (2007) The sky is your laboratory. Springer
Chromey FR (2010) To measure the sky. Cambridge University Press
Covington MA (1999) Astrophotography for the amateur. Cambridge University Press
Dragesco J (1995) High resolution astrophotography. Cambridge University Press
Dymock R (2010) Asteroids and dwarf planets and how to observe them. Springer
Hubbell GR (2012) Scientific astrophotography: how amateurs can generate and use professional imaging data, Springer
Smith GH, Ceragioli R, Berry R (2012) Telescopes, eyepieces and astrographs. Willmann-Bell

Websites

Cherry Mountain Observatory, www.cherrymountainobservatory.com
iTelescope, www.itelescope.net
LightBuckets, www.lightbuckets.com
New Mexico Skies, www.nmskies.com
Sierra Remote Observatories, www.sierra-remote.com
Sierra Stars Observatory Network (SSON), www.sierrastars.com
Slooh, main.slooh.com
University of Iowa Robotic Observatory (Rigel), astro.physics.uiowa.edu/rigel
University of Arizona Mt. Lemmon SkyCenter, Skycenter.arizona.edu
Warrumbungle Observatory, www.tenbyobservatory.com
Winer Observatory, www.winer.org
ADM Accessories, http://admaccessories.com/
Apogee Imaging Systems, http://www.ccd.com/
Astro Tech, https://www.astronomytechnologies.com/
Astro Hutech/BORG, http://www.sciencecenter.net/hutech/
Astro-Physics, Inc., http://www.astro-physics.com/
Atik Cameras, http://www.atik-usa.com/
Celestron, http://www.celestron.com/
Daystar Filters, http://www.daystarfilters.com/
Denkmeier Optical, Inc., http://www.deepskybinoviewer.com/
Explora-Dome (Poly Tank), http://www.exploradome.us/
Explore Scientific LLC, http://www.explorescientific.com/
Finger Lakes Instrumentation LLC, http://www.fli-cam.com/
Hotech Corp, http://www.hotechusa.com/
Howie Glatter's Laser Collimators, http://www.collimator.com/
Innovations Foresight LLC, http://www.innovationsforesight.com/
IOptron, http://www.ioptron.com/
Lunt Solar Systems LLC, http://www.luntsolarsystems.com/

Meade Instruments, http://www.meade.com/
Moonlite Telescope Accessories, http://www.focuser.com/
Peterson Engineering Corp, http://www.petersonengineering.com/sky/index.htm
Planewave Instruments, http://www.planewaveinstruments.com/
QHYCCD (Astro Factors/Deep Space Products), http://www.astrofactors.com
QSI, http://qsimaging.com/
Questar, http://www.questar-corp.com/
Santa Barbara Instrument Group, http://www.sbig.com/
Shelyak Instruments, http://pagesperso-orange.fr/shelyak/en/index.html
Sky Watcher, http://www.skywatcher.com/
Software Bisque, http://www.bisque.com/
Starlight Instruments LLC, http://www.starlightinstruments.com/
Starlight Xpress Ltd., http://www.sxccd.com/
Stellarvue, http://www.stellarvue.com/
Tele Vue Optics, http://www.televue.com/
Texas Nautical/Takahashi America, http://www.takahashiamerica.com/
Vernonscope, http://www.vernonscope.com/
Vixen Optics, http://www.vixenoptics.com/
William Optics, http://www.williamoptics.com/

Chapter 3

Remotely Located
Observatories Versus
Remotely Controlled
Observatories

A Family of Observatory Types

Today, several levels or types of observatories are available to you, the amateur astronomer. Most are designed to observe specific types of astronomical objects. Consequently, the first differentiator is object type. Deep sky objects are observed using long-exposure data acquisition equipment, whether it's with a charge-coupled device (CCD) imager or a spectrometer of some kind. Lunar and planetary objects are observed with a special type of digital video camera designed for that purpose (Fig. 3.1). Another way to differentiate these observatories is by the size of their astrograph's objective mirror or lens.

Most amateurs who own their own Astronomical Imaging System (AIS) have an astrograph that is up to 16 in. (0.4 m) in diameter. Most professionals, on the other hand, have access to larger AISs with astrographs that are 16 in. (0.4 m) to 336 in. (8.4 m) in diameter. Commercial services that provide access to AISs for professionals and amateurs alike provide astrographs that are in the transition zone—between 12 in. (0.3 m) and 32 in. (0.8 m) (Fig. 3.2).

Another way that observatories are different is their location and mode of operation. Before the turn of century, the classic amateur observatory was operated locally and for most amateurs, was a portable system. Since then, the hardware and software available to the amateur has enabled the automation of the instruments in the observatory and, most recently, enabled amateurs to configure their systems for remote control. Remotely controlled observatories can be locally located and remotely controlled, where the control system is within 100 ft (30 m); or remotely located and remotely controlled, where the AIS instrumentation can be located

© Springer International Publishing Switzerland 2015
G.R. Hubbell et al., *Remote Observatories for Amateur Astronomers*, The Patrick Moore Practical Astronomy Series, DOI 10.1007/978-3-319-21906-6_3

Fig. 3.1 Two types of imaging systems, deep sky camera and video camera

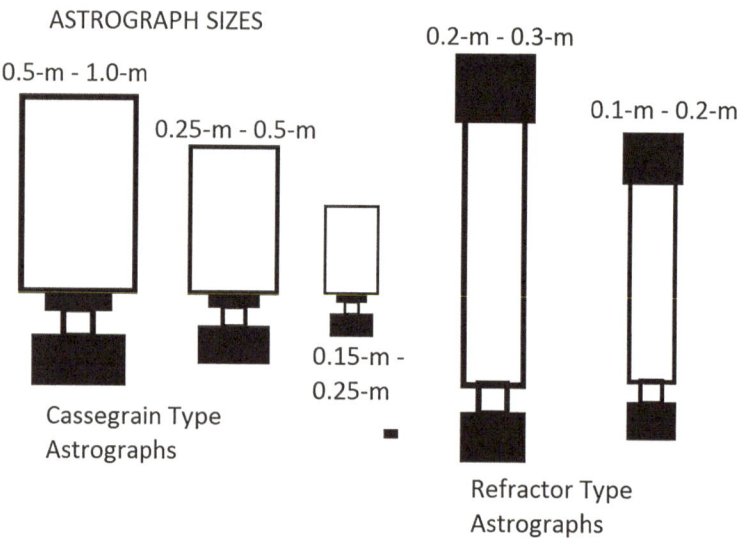

Fig. 3.2 Different astrograph sizes

anywhere in the world and operated from anywhere else in the world over the Internet from a distance of up to 12,000 miles (20,000 km).

So, if we were to list all the different types of observatories used today, they would make quite a list.

1. Small Astrograph, Local Control, Deep Sky Objects
2. Small Astrograph, Local Control, Lunar Planetary Objects

3. Small Astrograph, Remote Control, Deep Sky Objects
4. Small Astrograph, Remote Control, Lunar Planetary Objects
5. Large Astrograph, Local Control, Deep Sky Objects
6. Large Astrograph, Remote Control, Deep Sky Objects
7. Large Astrograph, Local Control, Lunar Planetary Objects

Many of these types of observatories are operated solely by amateurs, and others are operated by both amateurs and professionals. All have some common features as well as features customized for that particular observatory.

AIS enclosures also come in two types, the traditional domed observatory and a roll-off roof observatory. While the roll-off roof type is more simply constructed and is generally less expensive, some domed enclosures are cost competitive for smaller AIS systems while still conveying that traditional look and feel. Remotely operating a domed enclosure is also more complex than a roll-off roof observatory in that you must position the slit in real time as you track an object with your AIS.

The Base Design for a Remotely Controlled Observatory

All the observatories listed in the section "A Family of Observatory Types" have several characteristics in common in the design and in the equipment, although the performance expectation for some of the common components may be different for different observatories. An example is the required tracking accuracy for the mount is different for deep sky objects than for lunar and planetary objects. This is because the types of exposures are different for the two types of objects.

The base design requirements for an amateur-level observatory are also substantially different from the professional-level facility because of the huge difference in budgets available to build them. There are also differences in the equipment needed to outfit the two types of observatories for remote operations. Professional observatories have a wide range of systems and redundant equipment that the amateur observatory may not have, nor even require. That said, both professional and amateur remotely controlled observatories require a certain level of redundancy. The following is a basic equipment list for all observatories engaged in scientific deep sky imaging:

1. Astrograph and Mounting Plate
2. Astrograph Precision Mount
3. Mount Pier/Tripod
4. Guiding System Components

   (4a) Guide Scope
   (4b) Guiding Video Camera
   (4c) Guide Scope Mounting Bracket

5. Observatory Primary Power System
6. Observatory Backup Power Supply—Uninterruptible Power Supply (UPS)

7. Imaging Train Components

   (7a) Precision Powered Focuser
   (7b) Field Flattener/Focal Reducer Optical Element
   (7c) Photometric Filters
   (7d) Mechanical Adapters
   (7e) Deep Sky CCD Camera

8. Communications Components

   (8a) Ethernet Router—Redundant
   (8b) Powered Universal Serial Bus (USB) Hub—Redundant
   (8c) Powered USB Extension Cable 100 ft (30 m)
   (8d) Ethernet Cable
   (8e) Webcam

9. Control System Components

   (9a) Windows Based Computer System
   (9b) Data Storage Subsystem
   (9c) Weather Monitoring System
   (9d) Observatory Control Software
   (9e) Planetarium Software—Target Identification
   (9f) Guiding System Software
   (9g) Comprehensive Stellar Catalog—UCAC4
   (9h) Comprehensive Minor Planet Catalog—MPC
   (9i) Instrument Software Drivers
   (9j) Miscellaneous Control System Software

10. Observatory Enclosure

For a locally located, remotely controlled observatory that will operate as desired and expected, all of these items except #6, and the redundant items for #8a and #8b, are needed. However, these exceptions *are* needed for a remotely located observatory.

Remotely controlled observatories designed for lunar and planetary targets require other changes. These changes deal primarily with the imaging train components:

7. Imaging Train Components

   (7a) High-Speed Video CCD Camera

9. Control System Components

   (9a) Video Stacking Software—Registax

In addition, items #4 and #9f are not needed for this configuration.

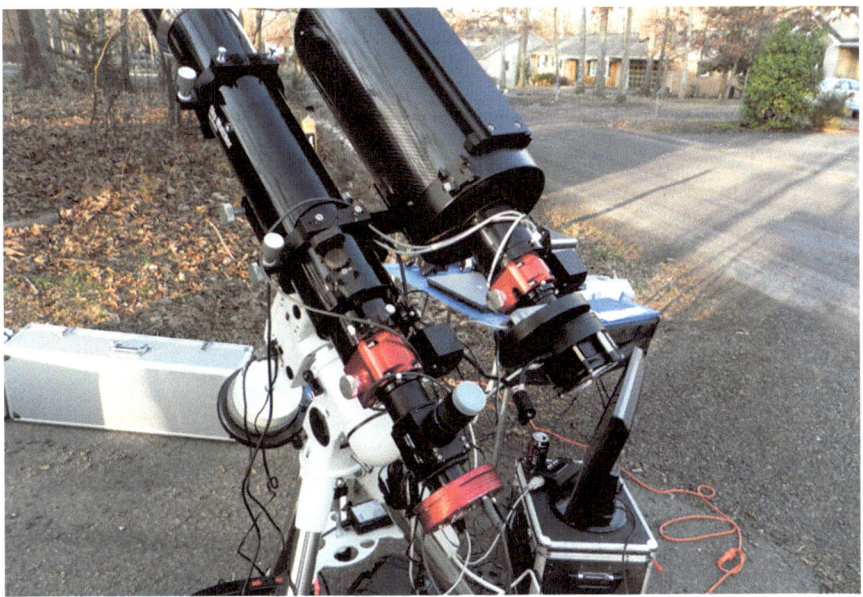

Fig. 3.3 Astrograph. Shown here is a refractor and a Cassegrain type astrograph

Astrograph and Mounting Plate

The astrograph is your primary instrument for gathering the light of any astronomical object you are studying (Fig. 3.3).

Because the primary function of the astrograph is to gather as much light as possible, you should strive to use one with as large an objective element as possible. In this way, you maximize the amount of light gathered and minimize your exposure time for a given magnitude for the objects under study. Different astrographs manufactured by various companies are delivered with other features depending on how much you want to spend. Some astrographs are delivered with manual precision focusers while others come without focusers so that you can provide your own powered precision focuser. Some astrographs have built-in fans to aid in equalizing the temperature of the primary mirror with the ambient temperature. This mitigates tube currents, providing an image affected only by the atmospheric seeing and not any local effects.

The astrograph should be as light as possible so that vibrations are damped out quickly. Most astrographs are available in carbon fiber to meet this goal. Carbon fiber is a very stable material and is not affected by temperature changes as an aluminum tube would be. This helps greatly in maintaining proper focus and minimizing the amount of time you need to spend focusing the instrument.

Astrograph Precision Mount

Because you want the best images possible from your system, you need a mount that performs well and reliably (Fig. 3.4).

In addition, your mount needs to be able to accurately track your objects with a low periodic error (less than 10 arc-seconds per worm period) and is otherwise smooth with no random spikes or excursions. These latter inaccuracies are the result of small defects and burrs or nicks in the mount's gears. It is common to use a guide camera and scope to correct for a smooth, continuously changing periodic error if these anomalous spikes are not present.

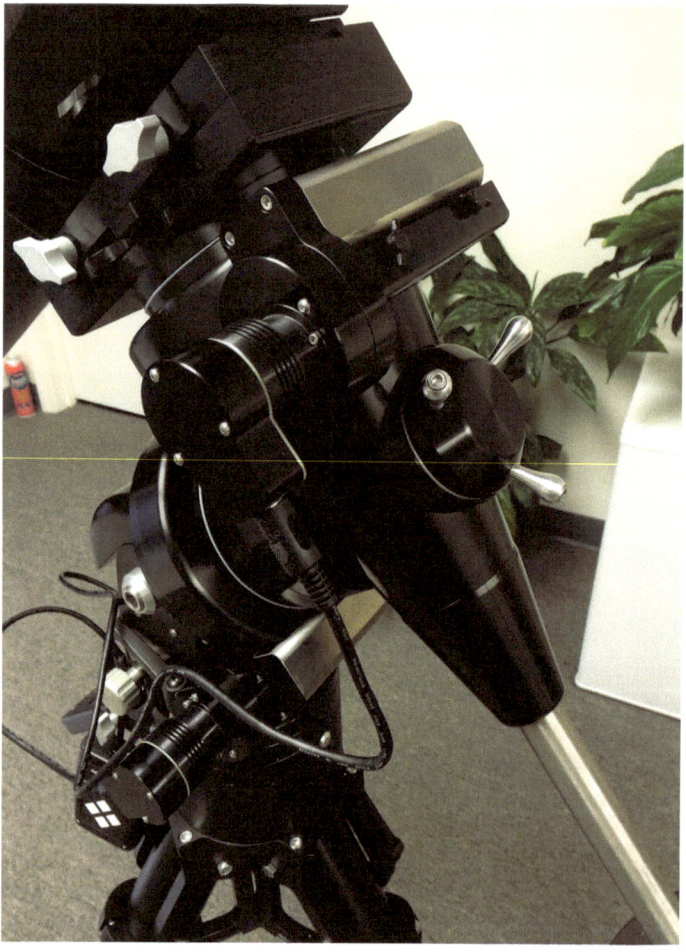

Fig. 3.4 Precision astrograph mount

Your mount's carrying capacity should be rated for about twice your expected load so that the gear system is not taxed and has an easy time tracking and slewing the astrograph, guide system, and imaging train combination. Again, this helps to maintain a smooth periodic error that is easy to correct with the guiding system.

Mount Pier/Tripod

The mount pier or tripod is essential in making sure that your telescope points consistently day in and day out (Fig. 3.5).

For a permanent observatory, a sturdy pier made of concrete and steel is a better choice than a tripod. While the pier can be anchored down, this is difficult to do effectively with a tripod. Tripods are designed to provide a stable platform for your mount in a portable form.

The observatory pier is usually made of concrete embedded in a concrete foundation a few feet underground below the frost line. The concrete comes up to the floor of the observatory, and then a steel pier is bolted to the concrete and leveled. The mount sits atop the pier and is finely adjusted for polar alignment. With fine-tuning, an excellent polar alignment can be achieved to the point where the declination drift is not apparent in your images even with exposures up to

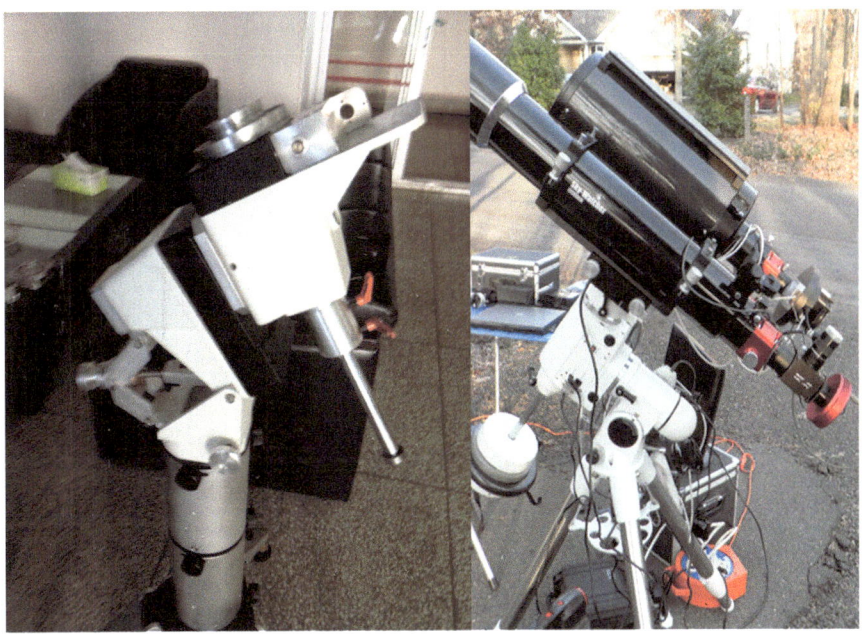

Fig. 3.5 Mount pier and tripod

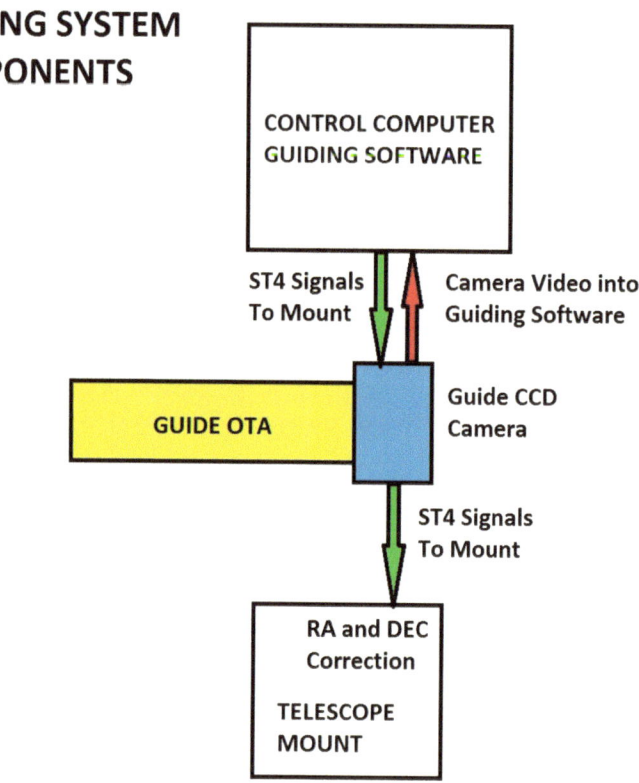

Fig. 3.6 Guiding system components

30–60 min. The only drift in the stars in your images at this point would be due to the inherent periodic error in your mount.

Guiding System Components

If you choose to install a guiding system on your AIS, it will allow you to correct for any small inherent periodic error caused by your mount (Fig. 3.6).

The guiding system consists of three main components—a guide scope, a digital video camera, and the guiding software. The guide camera software allows you to select a star in the field of view of the guide scope camera and then directs your mount through the guide input, also known as the ST4 connector, to make small corrections to the mount's RA and DEC position based on the star's position on the guide camera CCD.

The point is to keep the guide star fixed in the field of view so there is no relative motion between the sky and the telescope main camera. It is important to make sure that your guide system is mounted as sturdily as possible to minimize any possible flexure between the guide system and the main imaging system. You will see trailed stars if there is any flexure between the two systems. The typical correction rate for the guide system is .5 to 2 s. Unfortunately, if there are any rough spots on your mount's gear system (i.e., any burrs or nicks), then they may show up as small jumps in the position of the telescope. The guide system cannot correct these types of errors, and they will show up as small jumps in the star images in your photographs. Using a high-precision encoder system on the RA axis can minimize these types of errors.

Observatory Primary Power System

The primary power system for the observatory consists of the main AC power feed lines connected into a distribution panel that provides enough outputs to power each of the components of the observatory's AIS and support equipment (Fig. 3.7).

The distribution panel should be equipped with lightning protection and should have a robust ground provided by a ground rod adjacent to the observatory as shown in Fig. 3.7.

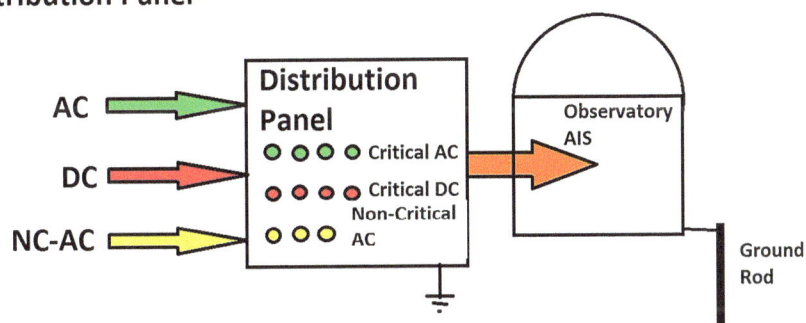

Observatory Power Distribution Panel

The Power Distribution Panel provides reliable connections for providing power to the AIS subsystems and components. It also provides power for other observatory systems including lighting and other facility purposes. It provides lightning protection also for the facility.

Fig. 3.7 Main power supply and distribution panel diagram

Observatory Backup Power Supply

The main power supply should be backed up via a backup system, which, depending on the load, can be a portable UPS powered by a battery system. Powered from the main AC supply, the UPS then provides power out to the distribution panel as shown in Fig. 3.7. The UPS should be able to, at a minimum, provide the full load to allow you to comfortably shut down the system without rushing, or ideally, provide the system with power to be able to finish the observing run (about 2–3 h).

The UPS should also provide lightning protection using metal oxide varistors or some other equivalent protection.

Imaging Train Components

Your imaging train will vary according to your observing program, but typically, several components will be common to most setups. Think of the imaging train as similar to a freight train. We start with the main engine and then couple several different train cars to make the whole train. The imaging train starts with the astrograph focuser (Fig. 3.8). The focuser is an important piece of equipment because it

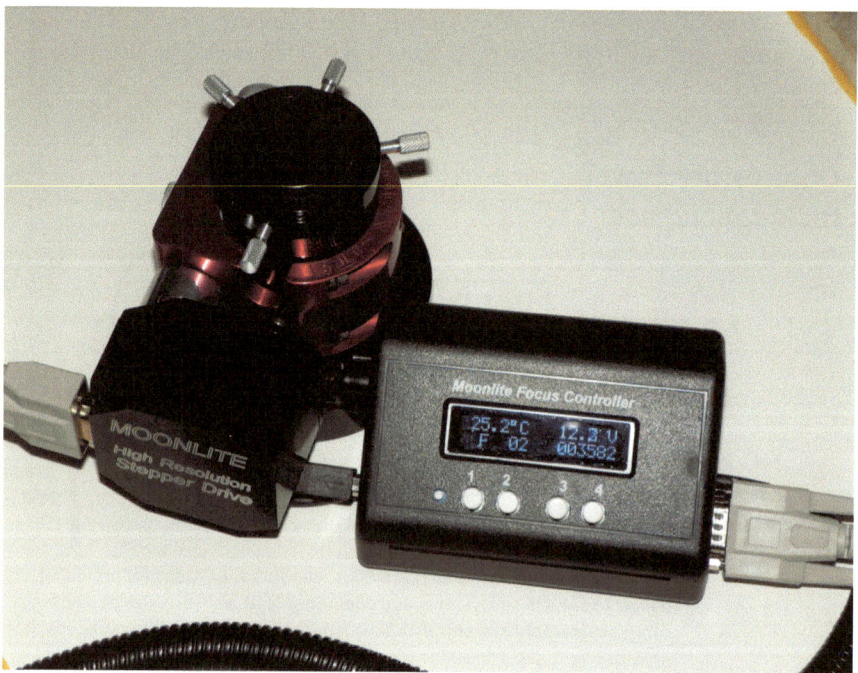

Fig. 3.8 Astrograph precision focuser

Fig. 3.9 Focal reducer and field flattener

allows you to precisely position your CCD imager at the exact focal point of your astrograph to within 10 μm (0.004 in.). The focuser should have a precision motor that allows you to keep track of its exact position.

The next item in your imaging train is typically some kind of field flattener and/or focal reducer (FR) (Fig. 3.9). This component corrects the image plane of your astrograph to make it as flat as possible. Also, the FR is used to match your image scale to the seeing conditions during your observations. It is important that you maintain a proper image scale when doing scientific imaging to be able to make the most accurate measurements from your images.

The next item in your imaging train is usually a spacer or coupler that properly connects and spaces the field flattener/FR from the next item in the imaging train (Fig. 3.10). You should always try to use a threaded coupler rather than a barrel type locking coupler because it is much more sturdy and rigid. A threaded coupler also does not sag like a barrel type of coupler with locking screws, keeping the imaging train square to the astrograph axis.

The next items in the imaging train are usually a filter or filter wheel system (Fig. 3.11). The filter wheel system should be electronically controlled so that you can remotely position a given filter along the imaging axis. A filter wheel can have from five to nine filter positions, depending on the model of filter wheel you choose to purchase.

Fig. 3.10 Standard 2-in. threaded coupler

Fig. 3.11 Filter wheel and filters

The final item that is typically a part of your imaging train is your CCD camera (Fig. 3.12). The camera should have a threaded connection to the filter/filter wheel system to ensure the camera is square with the rest of the imaging train and astrograph axis.

Figure 3.13 depicts a complete imaging train ready to mount onto the astrograph.

Fig. 3.12 CCD camera

Fig. 3.13 Complete imaging train ready to be mounted

The imaging train can be stored as a complete unit to save time when changing the system out in the field. When integrating the components of the imaging train, you will find that you need to do certain measurements to make sure that the focal point of the astrograph is placed on the CCD chip and that you have enough in- and out-focus range on your focuser.

Communications Components

Your remote observatory communications equipment plays a fundamental role in maintaining control over your system. Even though your AIS is a fairly benign object, it can be damaged by driving the mount into a position that causes your telescope to contact the pier or tripod on which it is mounted. Therefore, it is important either to make sure you can monitor your equipment in real time or to ensure you have set up the proper limits on mount position to absolutely prevent any contact with your pier or tripod. You need this latter level of confidence if you want to fully automate your system.

You need a reliable communications system, and there are a couple of things you can do to help to physically ensure that there will be no intermittent communications. First, use locking style connectors at both ends of any communications lines. Typically, your astronomical instruments, i.e., mount, camera, focuser, filter wheel, etc., use a USB style of connector. The cables from each of your instruments probably plug into some sort of USB hub that contains from four to seven ports. Routing your cables to avoid snagging or hanging when your focuser or mount moves goes a long way toward keeping things safe.

Knowing how your instruments respond to a command followed by a break in communications is necessary to ensure that no unsafe condition occurs. Test your mount by issuing a slew command and then breaking the communications link to see whether it completes the command successfully or relies on the communications link to complete the command. The controller should be able to complete the command on its own and place itself in a safe mode if the communications link has been removed. In addition, when your mount is tracking, it should have internal limits that you can define to ensure that it does not strike your pier or tripod. If it is a German equatorial mount (GEM), it should also execute the pier flip when its position passes the Meridian.

If you are using a USB hub as a concentrator, you need to ensure that the power connection is as secure as the port connections. Although a backup USB hub is probably not necessary, it is very convenient if your primary hub should fail for some reason and provides you with an opportunity to get back up and running in a short period of time. The primary USB cable that runs from your USB hub to your computer control system in your home or warm room should be run through a conduit if it is buried in the ground. If you are using an Ethernet connection to communicate with your equipment, you can provide a single network hub or a redundant hub, if you desire, to guard against a single hub failure. You should also install a lightning protection device or cable disconnect switch on this cable to protect the equipment from power surges.

Observatory Control System Components and Interfaces

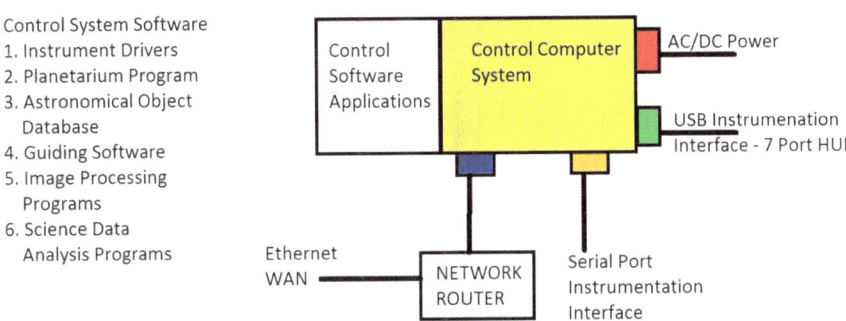

Control System Software
1. Instrument Drivers
2. Planetarium Program
3. Astronomical Object Database
4. Guiding Software
5. Image Processing Programs
6. Science Data Analysis Programs

Fig. 3.14 Observatory control system components

Control System Components

The remote observatory control system usually consists of a personal computer with the necessary USB or Ethernet interface. Your PC should have sufficient memory and hard drive space to support the numerous applications you plan to use for real-time data acquisition, data analysis, and reporting functions (Fig. 3.14). If you invest $1000 in a PC system today, then it should have the necessary CPU, memory, and disk space needed for your operation. As always, use your Observing Program Design Basis to define your specific needs.

The short list of applications you need to operate and maintain your remote observatory can be acquired commercially or may be available for free from various developers on the Internet. Several open-source applications are available with the source code included. This is very helpful if you want to develop a custom application or want to modify or add capabilities to existing open-source applications available to you. If you want to invest the funds in commercially available applications, you will find a good assortment of programs for any budget. You can purchase individual applications, or there are highly integrated applications that provide a range of services to operate your remote observatory, your AIS, and its full range of instruments. If you want to purchase commercially available programs, be prepared to spend upwards of $500–1000 to be fully equipped.

The following is the minimum list of applications, services, and functions you need to operate and maintain your remote observatory. (Several of these functions may be included in more than one application and also may be combined into a single application.)

1. Individual Instrument Software Drivers
2. Mount Control Software—Hand Controller

3. Focuser Control Software
4. Camera Control Software
5. Filter Wheel Control Software
6. Dome or Roll-Off Roof Control Software
7. Weather Monitoring Software
8. Guiding System Control Software
9. Planetarium Software—Electronic Star Charts
10. Observing Session Planning Software
11. Astrometric Star Catalog—UCAC 4
12. Minor Planet Catalog
13. Miscellaneous Astronomical Object Catalog
14. Mount Tuning Software—Periodic Error Correction
15. Astrometric Measurement Software
16. Photometric Measurement Software
17. Reporting Software—MPC Format, Other Formats
18. Internet Access software—Browsers, FTP Client, Email

Most of the items listed are integrated into a few programs that you can find on the Internet for free or are commercially available. Make sure that if the programs do any calculations that you are comfortable with the methodology and that you validate the results for yourself. Do not blindly accept the results without somehow verifying them until you are comfortable that the calculations are performed correctly. Providing bad results to professional groups that take your data is worse than providing no data. Several books are available to you that teach you how to perform data analysis and help you learn how to validate the results from your software.

Observatory Enclosure

The observatory enclosure provides several features needed to practically operate and maintain your AIS systems and subsystems. Chief among these is security. It is important that you are comfortable leaving your expensive equipment inside your observatory and away from your main dwelling. Securing your enclosure is a prime consideration when designing it, whether it's a roll-off roof observatory or a dome style. Consider carefully how to keep unwanted guests, such as burglars, insects, rodents, and other type of pests, out of your observatory. Securing any opening into your observatory may be as easy as a lock on a door, or it could be more complex, such as securing your roll-off roof so it cannot be lifted off to gain access. Sealing any openings will help keep the smaller pests out of your equipment.

The style of your observatory is a personal choice—you may like the traditional look and operation of a dome, or you may prefer the practicality of a roll-off roof observatory enclosure (Fig. 3.15).

Fig. 3.15 Domed roof enclosure for the Sierra Stars Observatory, Markleeville, CA

Either way, you need to plan your space to ensure that it meets certain criteria for the operational and maintenance needs of your observing program. The following are a few things you should consider when defining your space:

1. Determine whether you want to have only the equipment inside your observatory or whether you also want to be able to be inside yourself to operate it locally.
2. Determine whether you want to attach a "warm room" to your observatory from which you can operate the observatory for long periods of time in cold weather.
3. Provide space to locate all your electronic communications and other equipment above the floor and away from any openings to mitigate any direct access to the environment.
4. Make sure to provide a spot for a flat-frame target to take calibration flat frames.
5. Create a central location near your pier/tripod to place your UPS/power distribution panel.

If you choose a traditional domed enclosure, you will want to investigate all the manufacturers that provide domes. If you are very handy, you may want to construct your own dome. You can find several designs on the Internet from which to choose. A roll-off roof observatory enclosure can be a very cost-effective and efficient solution. You can find numerous design and construction plans on the Internet if you want to build from scratch, or you can modify a standard storage building available at your local home store. Either way, a roll-off roof enclosure is probably more practical in the long run, although it does not invoke the romantic image of the traditional astronomer within the domed telescope enclosure.

Additional Design Features for a Remotely Located Observatory

There are a few additional considerations when designing your remotely located observatory. Chief among these is the ground preparation for ensuring that your observatory has a long lifetime. Keep in mind that the most efficient enclosure enables you to equalize the inside and outside temperatures to bring your AIS into thermal equilibrium as quickly as possible. The materials used to build your enclosure should have high thermal resistance or low thermal conductivity and a low specific heat so the building doesn't store large amounts of thermal energy when exposed to the heat of the day.

Wood construction is common and is a good choice. Fiberglass is another option, especially when selecting a domed observatory. Although inexpensive, concrete block construction is not recommended because it has a large thermal mass that is not conducive to bringing your AIS to thermal equilibrium. The use of concrete should be limited to providing the mass for the telescope pier as a solid base for the mount. The floor of the observatory should be physically isolated from the concrete base to avoid inducing vibrations in your AIS when walking around the observatory. The observatory should be painted white or a very light color so it doesn't absorb energy and reflects light. Dark colors would exacerbate any thermal issues you may have with your observatory.

A roll-off roof observatory is very efficient for expelling the heat within the building once it is opened because of its large opening. A domed enclosure does not provide as large an opening to expel heat, so it may be necessary to provide a fan to help expel any heat that has built up in the enclosure when the dome shutter is closed.

Plan ahead to provide a conduit to route any cable systems you may run from the remote observatory to your base of operations in your home or other warm room. It's tempting to embed any conduit in the concrete pier base, but it may be better to just attach it to the outside of the base to facilitate any changes that may be necessary in the future. Make sure you seal any openings in the enclosure to mitigate infiltration of pests.

Operational Considerations for the Remotely Controlled Observatory

There are a few things to keep in mind when you operate your remotely located observatory. Use at least one webcam to view the movement of your AIS—it is very important to understand exactly what is going on inside your observatory.

A fixed view is fine but if you can afford it, you should consider a pan/tilt mount for your webcam so that you can view all around your observatory. You may also want to set up a game camera system to observe and record any critters that may find your observatory a nice home out of the cold.

Even if you have automated your observatory and made it possible to run it completely from your home, you may want to manually lock your enclosure, which will require you to physically visit your observatory prior to any observing session. In that way, you can do an inspection of your system and give yourself a warm, fuzzy feeling that everything is good, and you should not have any issue operating your observatory. You may want to write a pre-session checklist to verify that everything is okay before launching your observing session. Visiting your observatory before use can go a long way toward identifying wear and tear issues, or any pest-related issues.

Further Reading

Arditti D (2007) Setting-up a small observatory. Springer
Berry R, Burnell J (2005) The handbook of astronomical image processing. Willmann-Bell
Buchheim R (2007) The sky is your laboratory. Springer
Chromey FR (2010) To measure the sky. Cambridge University Press, Cambridge
Covington MA (1999) Astrophotography for the amateur. Cambridge
Dragesco J (1995) High resolution astrophotography. Cambridge University Press, Cambridge
Dymock R (2010) Asteroids and dwarf planets and how to observe them. Springer, New York
Hubbell GR (2012) Scientific Astrophotography: How Amateurs Can Generate and Use Professional Imaging Data, Springer, New York
Smith GH, Ceragioli R, Berry R (2012) Telescopes, eyepieces and astrographs. Willmann-Bell

Websites

Cherry Mountain Observatory, www.cherrymountainobservatory.com
iTelescope, www.itelescope.net
LightBuckets, www.lightbuckets.com
New Mexico Skies, www.nmskies.com
Sierra Remote Observatories, www.sierra-remote.com
Sierra Stars Observatory Network (SSON), www.sierrastars.com
Slooh, main.slooh.com
University of Iowa Robotic Observatory (Rigel), astro.physics.uiowa.edu/rigel
University of Arizona Mt. Lemmon SkyCenter, Skycenter.arizona.edu
Warrumbungle Observatory, www.tenbyobservatory.com
Winer Observatory, www.winer.org
ADM Accessories, http://admaccessories.com/
Apogee Imaging Systems, http://www.ccd.com/
Astro Tech, https://www.astronomytechnologies.com/
Astro Hutech/BORG, http://www.sciencecenter.net/hutech/
Astro-Physics, Inc., http://www.astro-physics.com/

Atik Cameras, http://www.atik-usa.com/
Celestron, http://www.celestron.com/
Daystar Filters, http://www.daystarfilters.com/
Denkmeier Optical, Inc., http://www.deepskybinoviewer.com/
Explora-Dome (Poly Tank), http://www.exploradome.us/
Explore Scientific LLC, http://www.explorescientific.com/
Finger Lakes Instrumentation LLC, http://www.fli-cam.com/
Hotech Corp, http://www.hotechusa.com/
Howie Glatter's Laser Collimators, http://www.collimator.com/
Innovations Foresight LLC, http://www.innovationsforesight.com/
IOptron, http://www.ioptron.com/
Lunt Solar Systems LLC, http://www.luntsolarsystems.com/
Meade Instruments, http://www.meade.com/
Moonlite Telescope Accessories, http://www.focuser.com/
Peterson Engineering Corp, http://www.petersonengineering.com/sky/index.htm
Planewave Instruments, http://www.planewaveinstruments.com/
QHYCCD (Astro Factors/Deep Space Products), http://www.astrofactors.com
QSI, http://qsimaging.com/
Questar, http://www.questar-corp.com/
Santa Barbara Instrument Group, http://www.sbig.com/
Shelyak Instruments, http://pagesperso-orange.fr/shelyak/en/index.html
Sky Watcher, http://www.skywatcher.com/
Software Bisque, http://www.bisque.com/
Starlight Instruments LLC, http://www.starlightinstruments.com/
Starlight Xpress Ltd., http://www.sxccd.com/
Stellarvue, http://www.stellarvue.com/
Tele Vue Optics, http://www.televue.com/
Texas Nautical/Takahashi America, http://www.takahashiamerica.com/
Vernonscope, http://www.vernonscope.com/
Vixen Optics, http://www.vixenoptics.com/
William Optics, http://www.williamoptics.com/

Chapter 4

Designing
the Amateur
Remotely Controlled
Observatory

Getting Down to Details—The Observing Program Design Basis

One of the best ways you can do yourself a favor and save some work when deciding how to build your remotely controlled observatory is to start by defining your observing program. This is done through the creation of your Observing Program Design Basis (OPDB). A properly documented OPDB will help you save money and time by focusing your attention on just those items necessary to achieve your observing goals. You can use your OPDB to define your equipment requirements and only spend your resources on those things that truly support your goals. Even if you are planning to use a commercially available Astronomical Imaging System (AIS), creating your OPDB will save you time in identifying those services that can provide the AIS that most closely matches your requirements. Once you have identified the observatory you want to use, you can use your OPDB to estimate the cost of imaging that will meet your observing goals.

The key to creating an effective OPDB is to understand your observing goals, specifically the types of measurements for the objects that you are interested in observing. Also you should fully understand the circumstances and the best conditions under which to observe your chosen objects, i.e., time of year, sky placement, seeing conditions.

Basically, there are three types of measurements you can do using a charge-coupled device (CCD) camera and other auxiliary imaging train components: astrometry, photometry, and spectrometry (Fig. 4.1). Astrometry is the precise measurement of astronomical positions in the sky, photometry is the precise

© Springer International Publishing Switzerland 2015
G.R. Hubbell et al., *Remote Observatories for Amateur Astronomers*, The Patrick Moore Practical Astronomy Series, DOI 10.1007/978-3-319-21906-6_4

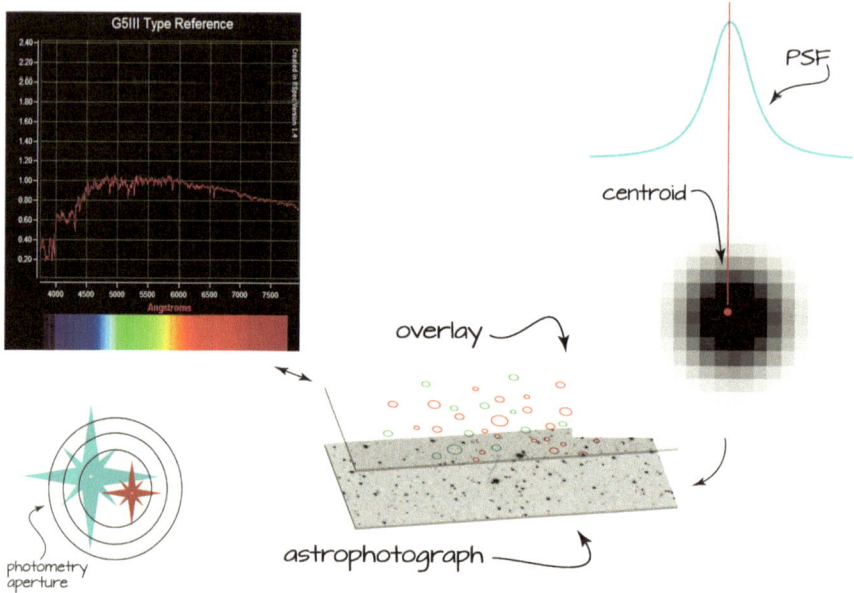

Fig. 4.1 The three basic astronomical measurement types: astrometry, photometry, and spectrometry

measurement of brightness of astronomical objects, and spectrometry is the measurement of the amount of light from an astronomical object over a given bandwidth with a given resolution. Each of these measurements requires a given set of conditions for seeing and magnitude, depending on the light-gathering capability of your astrograph. Table 4.1 outlines the information you should consider when developing your OPDB.

As you can see, the items listed Table 4.1 apply directly to creating a Minimum Equipment List (MEL) (Fig. 4.2) that will satisfy each of the items. The items also drive the procedures you need to develop to operate your AIS efficiently and obtain the best data possible. That should be your primary goal when developing and using your procedures. The limits you place on the performance of your AIS will drive the total cost of the systems, subsystems, and components (SSC) you decide to purchase and integrate into your AIS. To minimize the total cost of your AIS, every SSC should exist to meet one or more of the listed requirements. You should never have an SSC that is not needed to meet one of your listed items. That is why your OPDB should drive your MEL; it is a waste of money and time to support unneeded components.

Just as important to the safe and consistent operation and maintenance of your observatory is the use of procedures (Fig. 4.3). The procedures are generally of three different types, AIS Maintenance, AIS Operations, and Data Processing.

Table 4.1 OPDB parameters and example values

Item	Parameter	Example values
1	Observing Goal	Measure the Light-Curve of a Minor Planet
2	Desired Results of Measurement	Determine the Rotation Rate of Minor Planet, Determine the Type of Supernova Observed
3	Observing Location	Latitude, Longitude, Altitude, Address
4	Observing Time of Year	Spring, Autumn, January–March
5	Object Type	Deep Sky, Stellar, Planetary, Minor Planet, Supernova, Binary Star
6	Object Name	Binary Star 62 Cygni, Minor Planet 4150 Starr
7	Object Magnitude Limit	>12th mag
8	Measurement Type	Astrometric, Photometric, Spectroscopic
9	Calibration Frames Required	Dark, Bias, Flat
10	Astrograph Type	Refractor, Cassegrain, Ritchey-Chretien (RC)
11	Astrograph Size	0.3 m, 0.15 m, 8 in.
12	CCD Image Scale per Pixel	1 arc-second, 1.5–2.5 arc-seconds, <2 arc-seconds
13	Mount Type	Fork, German Equatorial, Alt-Az
14	Is Mount Permanently Installed?	Yes/No
15	Is AIS Remotely Controlled?	Yes/No
16	Is AIS Remotely Located?	Yes/No
17	Level of AIS Automation	Fully, Partially Unmanned, Partially Manned, No Automation—Manually Operated
18	Redundancy Level	Fully, Partially—Communications, Partially—Instrumentation, Partially—Power Systems, No Redundancy
19	Maximum Acceptable Seeing (MAS)	<3 arc-seconds
20	Minimum Acceptable Signal to Noise Ratio (MASNR)	>75, >100
21	Object Sky Location Limits	Near Meridian, Near Horizon
22	Frequency of Observation Sessions	Twice a Month, Three Times a Week
23	Photometric Passband	V-band, UVBRI, Infrared, Not Applicable
24	Spectroscopic Passband	4200–5100 Å, 380 nm–820 nm, Not Applicable
25	Image Exposure Time	60 s, 180 s
26	Maximum Acceptable Periodic Error (MAPE)	<5 arc-seconds
27	Minimum Length of Observing Session	3 h, 4 h
28	Maximum Length of Observing Session	8 h, 12 h
29	Environmental Limits	Minimum Temperature = 30 °F, Maximum Cloud Cover = 10 %
30	Other Notes As Needed	Startup Date 01 March 2017, Required Procedures

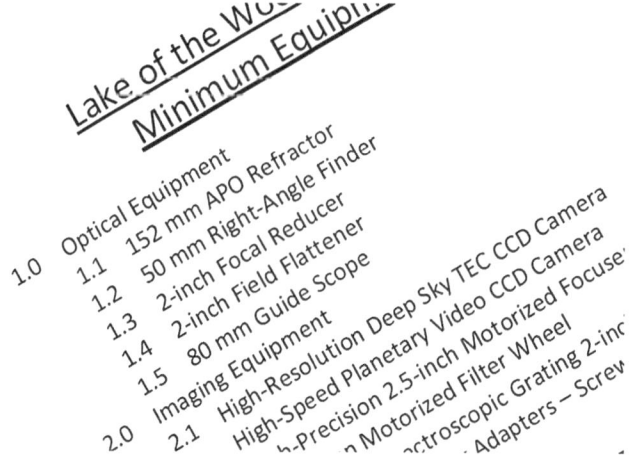

Fig. 4.2 Minimum equipment list

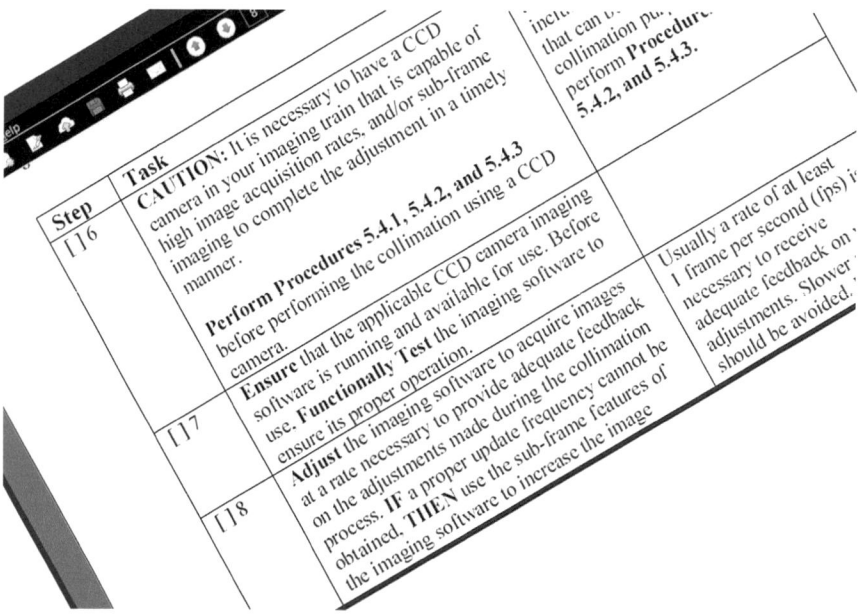

Fig. 4.3 An AIS procedure

Table 4.2 outlines the various procedures you may need to support your AIS Operations and Maintenance (O&M) and states the purpose of each procedure. (See Chap. 5 for examples of some of these procedures for your use.)

The procedures listed Table 4.2 are by no means all there could be. If you identify a special need to record or document a configuration or operation, then by all

Table 4.2 Example procedures

Item	Procedure name	Procedure purpose/description
1	AIS Equipment Checklist	Check condition of equipment
2	AIS Equipment Integration	Complete Initial System Integration Checklist
3	AIS Imaging Train Configuration	Calculate factors and assemble the imaging train
4	AIS Spectroscopic Equipment Configuration	Calculate factors and assemble the spectroscopic system
5	AIS Equipment Power Up	Power up AIS SSCs in correct sequence
6	AIS Communications Verification	Ensure that SSCs are communicating successfully
7	Acquire Imaging System Calibration Data	Acquire dark, bias, and flat frames
8	AIS Mount Polar Alignment	Align the mount to the celestial north pole using a drift alignment or other applicable technique
9	AIS Sky Alignment	Perform a one-star, two-star, or multi-star sky alignment and verify mount pointing performance
10	Calculate Astronomical Object Parameters	Calculate the celestial location, imaging train exposure information, and other values for the astronomical object under study
11	Data Acquisition	Perform steps to acquire desired object data
12	Calibrate Image	Apply calibration frames to object data (light frame)
13	Astrometric Measurement	Perform a plate solve on the calibrated light frame, measure object position
14	Photometric Measurement	Perform a photometric measurement on desired objects in frame, calculate object magnitude to desired SNR
15	Measurement Reporting	Create results report in required format according to standards of recipient organization, e.g., Minor Planet Center (MPC), Association of Lunar and Planetary Observers (ALPO), American Association of Variable Star Observers (AAVSO)
16	AIS System Shutdown	Perform a controlled shutdown of your AIS
17	AIS Equipment Storage	Define how your AIS system components are to be kept in long-term storage

means create a procedure for it if it is something that you will be doing in the future on a regular basis. It is important to understand the reason for and importance of procedures, and how they contribute to your overall success in meeting your observing goals.

Procedures are designed to guarantee consistent AIS performance and operations and minimize the time and effort in running your observatory. First of all, your memory is not perfect—no matter how much you practice and learn a specific operation, you can forget something. Alternatively, even if you do not forget anything during a given operation of your equipment, a procedure can be needed to document results and help to identify operational glitches that you may not immediately recognize. Procedures always help you maintain consistent results. Procedures are great for making sure you perform a series of steps in a specific order so you don't damage your equipment or forget to record specific data of a time-sensitive nature.

The initial creation of your procedures will be based not only on the training and experience you have obtained over the years but also on the SSC manufacturer's owner's manual and other materials provided (Fig. 4.4). The manufacturer's website is another source for additional information pertaining to the operation and maintenance of their equipment. However, the owner's manual is the primary source for the operation of your equipment—make sure you stick to it in case you run into problems that require any warranty work. Also be aware that if you have tried to operate your equipment outside the limits set by the manufacturer, you will have a difficult time getting it fixed if problems occur.

During your initial AIS system integration when you are getting all your parts together and building your system, the MEL and Initial Integration procedure help you to build your system in a specific sequence that minimizes any inadvertent damage you may cause owing to your unfamiliarity with the components. Typically, the most likely time for damage to occur is when you are putting disparate components together using your initial design for coupling these components. Finely threaded coupling components can be damaged if you run into a part that has not be cleaned or de-burred properly. This type of defect can be easily overlooked.

The other purpose for procedures is they allow you to record session-specific information dealing with equipment configurations, settings, parameters, and other data that you cannot rely on your memory to retain. The parameters and other configuration data you record during your observing session helps you to improve your operations going forward. Procedures also help you in providing evidence of your configuration if someone were to question how you obtained your data or the validity of the data. You can demonstrate conclusively how your data were gathered and, if necessary, identify any systematic errors that may have been introduced into your system to cause a problem. Providing documentation on the O&M of your AIS demonstrates your professional approach to gathering your data and lends a lot of credibility to your operations.

Finally, procedures will help you in the future when changing your configuration or adding new equipment to your AIS. One important task you will want to perform is called regression testing. In regression testing, you retest previously configured equipment to which changes were made to ensure that you have not inadvertently broken anything or adversely affected the performance of your SSCs. A typical example may be when you receive an updated software driver for your main CCD camera. You want to test the new driver in the configuration you typically use your camera, i.e., using exposure times and other specific CCD parameters. Ensure that you verify that any calibration frames you take are not also adversely affected by the updated driver.

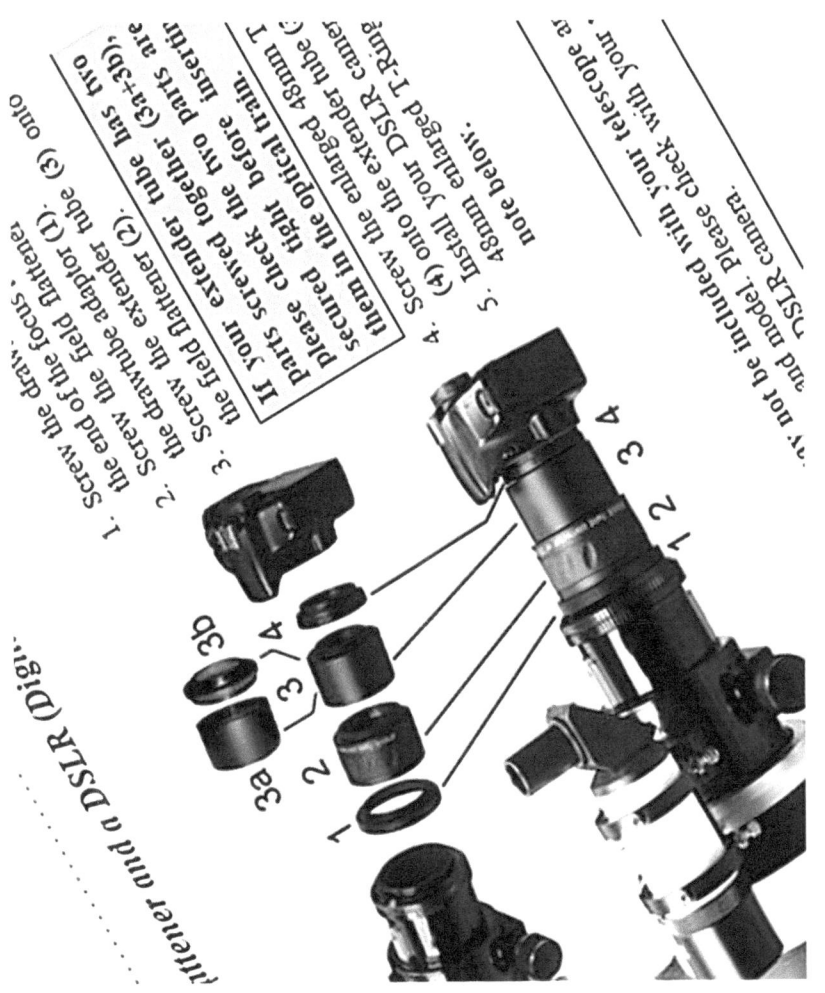

Fig. 4.4 The manufacturer's owner's manual is a prime source of information for creating procedures

Another example may be when you add additional equipment to your AIS that may affect the weight and balance of your AIS. You need to make sure that the dynamic response of your mount's drive performance is not adversely affected by the added weight (inertial load) causing poor tracking performance. This test is paramount in maintaining excellent imaging results. You may also see an impact on your mount's pointing performance as a result of the added weight.

The Base Astronomical Imaging System (AIS)

The MEL for the base AIS is designed to list those items that directly affect the performance of the AIS in meeting your OPDB requirements. The base MEL contains the core of your AIS and will be common to any type of observatory you may build, whether it is portable or permanent (Fig. 4.5). Using the OPDB parameters listed in Table 4.1, address each of them as best you can as an initial start of the process. As you go about identifying your SSCs, you will refine your OPDB to more closely match your goals and the equipment that is available and affordable. If you plan to use a commercial service, then you can adjust your OPDB to more closely match the capabilities of the observatories to which you have access. Table 4.3 shows an example OPDB.

The example OPDB shown in Table 4.3 describes a project to measure the rotation rate of a minor planet. This example is typical of projects that amateurs and professionals alike engage in. The International Astronomical Union (IAU) MPC is the foremost authority on minor planets and is the clearinghouse for all data regarding minor planets. It collects astrometric and photometric observations of minor

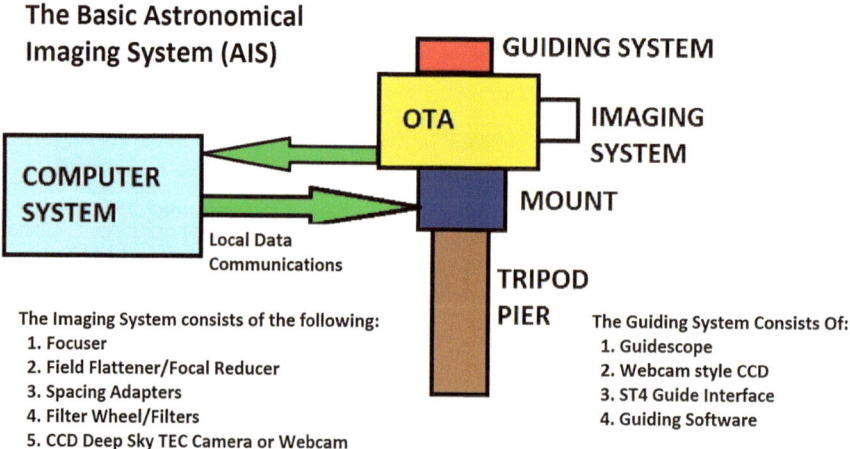

Fig. 4.5 The base Astronomical Imaging System (AIS)

Table 4.3 Example OPDB

Item #	Parameter	Value
1	Observing Goal	Acquire data to create a light-curve of a minor planet
2	Desired Results of Measurement	Measure the rotation rate of a minor planet
3	Observing Location	38d 20m N 77d 46m W MPC I24
4	Observing Time of Year	All year
5	Object Type	Minor Planet
6	Object Name	Any Main Belt Asteroid
7	Object Magnitude Limit	>12th Magnitude
8	Measurement Type	Astrometric, Photometric
9	Calibration Frames Required	Bias, Dark, Flat
10	Astrograph Type	Apochromatic Refractor
11	Astrograph Size	0.152-m f/7.5
12	CCD Image Scale	1 arc-second/pixel
13	Mount Type	German Equatorial (GEM)
14	Is Mount Permanently Installed?	Yes
15	Is AIS Remotely Controlled?	Yes
16	Is AIS Remotely Located?	No
17	Level of AIS Automation	Manual Operation
18	Redundancy Level	Partial-Communications, Partial-Power Supply
19	Maximum Acceptable Seeing (MAS)	4 arc-seconds
20	Minimum Acceptable Signal to Noise Ratio (MASNR)	>75
21	Object Sky Location Limits	Near Meridian
22	Frequency of Observation Sessions	3 Times Per Week
23	Photometric Passband	V-Band Photometric
24	Spectroscopic Passband	Not Applicable
25	Image Exposure Time	60 s
26	Maximum Acceptable Periodic Error (MAPE)	<2 arc-seconds With High-Precision Encoder Drive Corrector
27	Minimum Length of Observing Session	4 h
28	Maximum Length of Observing Session	8 h
29	Environmental Limits	Partly Cloudy, >20 °F
30	Other Notes As Needed	None

planets and performs all the orbital calculations needed to keep track of these small bodies. The ALPO Minor Planet Section publishes a monthly newsletter called *The Minor Planet Bulletin*. It contains observation reports from observatories all over the world that contain light-curve and calculated rotation rates of minor planets.

The AIS that meets the requirements of the example OPDB can be described as follows:

The AIS consists of a 152-mm Apochromatic (APO) Refractor with a focal ratio of f/7.5. It is mounted on a precision German Equatorial Mount (GEM) capable of tracking with a *periodic error (PE) of <2 arc-seconds. This is accomplished through the use of a high-resolution encoder drive corrector. Because the required exposure time is 60 seconds, this system does not require a guiding system. The imaging train consists of a precision motor-ized focuser, a flat-field correction optic, a V-band photometric filter, and a Kodak 8300 based, 8.3 megapixel CCD camera that is thermally cooled down to 40 deg C below ambi-ent temperature. A filter wheel system is included that contains the V-band filter. The AIS is controlled remotely via a universal serial bus (USB) communications system consisting of a powered seven-port USB hub and a powered 30-meter extension cable. The control soft-ware runs on a laptop personal computer (PC) and consists of MaximDL and the EQMOD ASCOM driver system. The system is aligned and pointed via the Carte du Ciel (CDC) freeware planetarium program. The plate-solving software Pinpoint is integrated into MaximDL and is used to calculate the minor planet astrometric position. The program Astrometrica is also used to calculate the astrometric position of the asteroid and create the MPC report file. Both Pinpoint and Astrometrica use the USNO UCAC4 Catalog. This cata-log provides all-sky high-precision stellar positions for stars up to 18th magnitude. It also provides proper motion values for the majority of those stars for accurate positions down to less than 0.1 arc-second. Good photometric values in the V-band are also provided in this catalog. This system is capable of imaging 12th magnitude asteroids with an SNR of at least 75. This system is located in a rural area that has fairly dark skies and good seeing, averag-ing 2–3 arc-seconds. In the summer, the seeing has been observed to be a consistent 2 arc-seconds or less. The location has been assigned the Minor Planet Center observatory code I24 Lake of the Woods Observatory, Locust Grove, VA. This system is operated manually with the PC control computer located next to the AIS instrumentation.*

The AIS described above costs approximately $12,000–15,000 USD and will provide excellent results for this observing program. This section has described the base AIS with the exception of the powered USB extension cable. As stated previ-ously, the base AIS provides all the primary equipment needed to accomplish the goal of the observing program. Other subsystems are needed to turn this base AIS into a remotely controlled and a remotely located observatory. These are described in the following sections.

SSCs for Remotely Controlled AISs

The requirements of a remotely controlled AIS add a few items to the MEL of the base AIS described in the section "The Base Astronomical Imaging System (AIS)." It is fairly straightforward to add remote control capability to a locally located AIS. Because the control system is located within 100 ft of the AIS, it is simply a matter of providing a high-speed communications cable between the AIS instru-mentation and the control system computer, which, in most cases, is a laptop with some kind of external high-capacity storage subsystem. In the example OPDB, a 30-m powered USB extension cable (Fig. 4.6) is used to connect a seven-port USB hub, which is located between the AIS and the computer system. At a minimum, this is all that is required to turn a standard AIS that is manned into a remotely controlled AIS. There is the requirement to provide a power source at the AIS instruments that is separate from the power source for the control system computer, because the latter is more than likely located at a home office or other warm space.

Fig. 4.6 A 30-m powered USB extension cable

If it is a longer distance between the AIS instruments and the control system computer, then some type of USB serial-to-Ethernet converter and an Ethernet router are required. Ethernet can be used to bridge a distance of several hundred feet if necessary. If you decide to use Ethernet, you can also provide for failures by including redundant Ethernet routers between the AIS and the control system computer. You also need to provide a length of power cable between your main AC power location and your AIS observatory. Providing an uninterruptible power supply (UPS) at the AIS observatory location would also allow a certain level of protection in case of power outages. At a minimum, this would allow you to shut down your system and/or finish up an observing session before using up all the backup battery capacity available.

Additional SSCs for Remotely Located Observatories

The challenge of designing and building a remotely located observatory cannot be overstated. The fact that you are at least several if not hundreds of miles away from the observatory means that, at best, you can see and operate your AIS, and at worst, you cannot communicate at all with your system. The observatory might as well be in Earth orbit for all you will be able to do in the immediate future to resolve a problem. That is unless you have equipped your observatory with a few redundant items and provided a means for you to maintain communications with your AIS and perhaps even correct any issues you may experience with operating your system.

There are many ways to accomplish the goal of building a highly efficient remote controlled observatory. A highly reliable system provides protections against failure at different levels in the system. Also, a highly reliable system will fail gracefully, a small portion at a time, rather than failing in a catastrophic way (i.e., everything is 100 % operational one moment and then in the next moment completely fails, with no systems working). Approaching the design of the system from both top down and bottom up can allow you to include inherently redundant design features and provide the graceful failure modes that you desire in a system.

Within the SSC hierarchy, starting with the component level, some components are more important than others; these cannot be allowed to become a single point of failure. These single-point vulnerabilities can be the bane of an otherwise reliable design. In addition, the built-in redundancy at the component level can be part of the hot standby configuration or of the cold standby configuration. For example, the motor controller in the telescope mount can provide single-train motor drivers for each of the axes so that if the motor driver chip fails, then that axis fails to operate. The controller could be designed to provide two motor driver chips that are connected such that one is the primary driver and the other is the secondary driver. They are connected such that if the primary driver chip does not produce any signals, the secondary driver chip's signals are provided to the motor through an auctioneering circuit. This example would be implemented on the controller board itself. An alternative configuration is to provide two completely separate controller boards that are connected together such that they receive the commands simultaneously but only the primary controller acts on them. The secondary controller acts on the commands only if it detects that the primary controller has failed to act. Again, this would be implemented using an auctioneering circuit through which the primary controller would listen for completed actions. These types of design features can eliminate single-point vulnerabilities.

Another approach to mitigating failures that would cause a catastrophic shutdown of the system addresses the reliance of all the AIS subsystems on a common power source. If that power source fails, then everything would shut down. This type of failure is called a common mode failure. One way around this situation is to provide each subsystem with an independent source of power. Common mode failures can be obvious or insidious. Here are two examples.

The first example involves the power system that has been specified. If you have specified a type of lightning protection circuit for each of your cable systems and power inputs that uses the same hardware, and the hardware has a substandard component that is prone to early failure, then all the systems protected by that hardware have a common mode failure potential that could shut down your system if a single storm came through and caused the protection to fail.

The second case is a more insidious form of common mode failure. For example, you have provided redundant controllers configured as previously discussed to guard against a single controller failure. In this case, however, the failure mechanism is actually within the firmware installed on the controller. A bug has been introduced into the firmware such that a specific sequence of events places the motor driver in an unsafe mode. Because both controllers have identical firmware, they behave in the same way when presented with the same command sequence. Although you have provided redundant hardware, the firmware is common and so acts as a single-point vulnerability. In this case, and in the one described previously, you can see that providing a redundant device has not protected you from a common mode failure that results in a single-point vulnerability.

The concept of *defense-in-depth* covers the redundancy requirements, but an additional concept called *diversity* addresses the common mode failure issue. The overall concept is called *defense-in-depth and diversity* (DIDD) (Fig. 4.7). In the second example above, the common mode failure could just as easily have been related to a flaw in the hardware components on the controller board because they also use identical hardware; if there is a flaw in the hardware or in the circuit design, it could easily lead to a common mode failure. In the second failure example described above, the firmware had a bug that presented during a specific command sequence. Using the diversity philosophy, you would have provided the two controllers, each with a different firmware program that accomplishes the same

Defense-In-Depth and Diversity

FUNCTION 1()

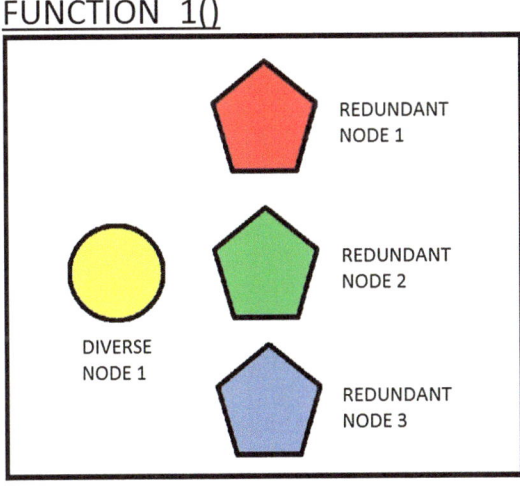

Redundant Nodes 1, 2, and 3, and Diverse Node 1 all perform the same function. The redundant nodes provide 3 channels of backup in case of a single point failure. If there is a common mode failure in each of the redundant nodes, then Diverse Node 1 provides service to ensure continued operation

Fig. 4.7 Defense-in-depth and diversity

controller functions and behavior, but that do not have any code in common. That way, if one of the controllers fails on a specific command sequence, the other should not because it is not executing the same identical code to accomplish that command sequence.

Digital systems are prone to this type of common mode failure, and in the nuclear industry, the diversity concept is used to guard against these failures. One way to handle the issue is to provide two separate but equal systems. One is of digital design with its attendant redundant components, and the other, completely separate system is an analog system providing the same functions and performance as the digital system. The analog system is completely foreign compared with the digital system, and in no way can fail in the same way the digital system would fail. It's as if a digital alien visited an analog planet. The digital alien would not speak the same language, would not even have any identifiably similar organs as the analog beings on the analog planet. They could not interface with each other, nor combine components of each other's bodies and expect them to work. Viruses that affected the analog beings would be harmless to the digital alien, and vice versa. This is the strength of diversity in the face of common mode failures.

Power systems are an important case because providing DIDD protections can be a little challenging and taken to the ultimate, can be quite expensive. Now, if you were designing an orbital satellite system you would, of course, spare no expense in providing for DIDD through the specification of a group of power systems each carrying its share of the load. Ground-based observatories are in a different league. Unless you are observing a once-in-a-lifetime celestial event, there is some tolerance for failures of your observatory. It is not necessary to provide the expensive diversity of primary power sources, which is typically an AC main power supply from your local utility. Consider the expense of having two utilities that are completely separate from each other providing power lines to your remote observatory location. The expense would be extreme if it were even possible. This would be overkill. Providing a simple battery UPS solution on top of the main AC power feed is typically more than enough to allow time to safely shut down your observatory AIS and close up shop for the night.

Communications systems are similar to power systems in that although they are important, simply adding basic redundancy features will enable you to maintain communications to safely shut down your AIS (Fig. 4.8). Because you may be several to hundreds of miles away from your AIS, it goes without saying that you will be relying on a high-speed Internet connection for your main communications needs. It still is possible to provide a slow speed connection over a dial-up modem system, but this would only be sufficient to perform the most basic functions, such as power down and power up your remote systems, if your Internet connection were not available. Because this may be all it takes to recover a lost Internet communications link, it may be worth the effort to install a dial-up capability. A cellular telephone system may be all it takes to provide this basic power sequencing system.

It won't be the end of the world if you lose communications to your observatory, as long as the local controllers know what to do when they lose touch with you. The default behavior should place your systems into a "safe" mode so that they either immediately stop what they are doing so they don't cause any unmonitored harm,

REDUNDANT
COMMUNICATIONS

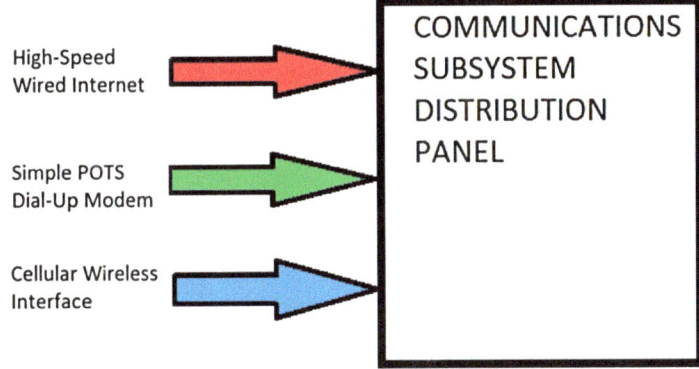

Fig. 4.8 Redundant communications features

or they have the smarts to complete their last command and then place themselves in a safe mode. The only real danger if you lose contact with a remote observatory is if there is an approaching storm and the observatory roll-off roof does not close to keep your equipment from getting wet. Not to downplay that possibility…it can be quite devastating.

You need to understand these concepts not to design your own components, but to understand the possible failure modes of your equipment based on the information the manufacturers provide you about the design and operation of your equipment. Advanced concepts such as DIDD are designed into systems that are becoming affordable for the lower budget, professional-level systems. You can count on having these features in the lowest cost systems in the next 10–15 years as the technology becomes cheaper to implement and the hardware costs continue to drop. By educating yourself and asking the manufacturers to provide these DIDD features, you help to usher in the next level of performance and reliability for your remotely controlled, remotely located observatory. The end result will be smooth operations and few to no failures that make your equipment "disappear" so that all you see and care about is your data stream from that far-flung astronomical object that you study day in and day out.

Design for Operations—The Importance of "Not Being There"

The concept of Design for Operations (DfO) gets to the heart of building an efficient remotely operated AIS and thus minimizing your physical interaction with the system to accomplish your OPDB goals. The bulk of the performance requirements to gather data of sufficient quality on a consistent basis is accomplished through the identification of specific equipment. This equipment need not be "over qualified" (so-to-speak), but only needs to consistently meet the performance requirements. It is easy to over-buy equipment and waste money. Using only the minimum equipment necessary to meet or exceed your OPDB goals means no wasted effort, time, or money. You need to read and heed all the advice in the section "Additional SSCs for Remotely Located Observatories" so that you can intelligently provide only the necessary level of DIDD to minimize the cost and maximize the reliability of your AIS and remote observatory. Carefully read all the manufacturers' literature on the O&M of their equipment and assess the cost-benefit ratio. The concept of "keep it simple and stupid" (KISS) can apply in this context because you minimize the amount of equipment needed to do the job effectively. At first, this may seem to contradict the need to provide DIDD features, but this isn't so. The goal is to provide only the minimum DIDD features necessary to maximize the reliability and minimize the cost. You need to apply a balanced approach to design the most efficient configuration of AIS equipment and DIDD components to maximize the value of the system.

One of the decisions you will make when designing and building your AIS and observatory is how much automation you want to include. A fully automated system similar to those commercial services to which you submit a schedule and the schedules run overnight while you sleep takes a lot of work to perfect. However, typically, CCD camera control software only allows you to take a series of exposures with a given set of parameters on one target at a time. This would entail you manually slewing to a given target location and then starting the scheduled exposures. If all you are interested in is observing one target and you have a series of exposures scheduled that would take several hours, as long as your mount's tracking is accurate, you can probably retire to bed and set your alarm to wake you when the exposures are complete. This is an easily obtainable level of partial automation. Additional software is available to accomplish full automation, and while it is somewhat cost prohibitive on an amateur level, it is absolutely doable if you so desire. Fully automating your system will afford you the opportunity to start your system up, sleep the night, and then shut down your system in the morning fully refreshed and ready to process your data.

Because your observatory is, by definition remote and unmanned, you will need to consider its physical security. When your AIS is in operation, the roll-off roof or the shutter of your dome is necessarily open, and therefore vulnerable to anyone who may decide he/she wants your equipment. You need to seriously consider providing some kind of silent alarm to your location to warn you of any unexpected movement inside your observatory. Typically, if anyone gets inside your observatory while it is operating and starts to take things apart, you will know it if you run your system manually. If your system is automated, it should be equipped to wake you up if operations have ceased for some reason, and it may be worth the investment to provide a commercial security system that will also notify authorities on your behalf. Having some kind of trip wire system can accomplish this goal. Finding out that your observatory was burglarized after the fact is not a good option; it's better if you can catch the person in the act and dispatch the authorities immediately.

You are making the effort to do a thorough design so that you can be comfortable running your observatory from a warm location away from your AIS. Not being there is your reward for designing a solid AIS and observatory. Your work will reward you with trouble-free operations and high data throughput as you slew your system from target to target gathering your data. Apart from the pride of ownership of your observatory, you can be proud of the professional results you can achieve through your attention to detail and perseverance in completing the task of designing and building a world-class observatory capable of providing years of service.

Design for Maintenance—The Importance of "Not Having to Go There"

In designing the remote observatory to minimize visits to the site, you have necessarily minimized the amount of equipment and components needed to do the job at hand. This is called Design for Maintenance (DfM). The less you have, the less there is to go wrong. In addition, you need to consider those times when you have failures. First, you should be able to recover from any failure you encounter without going to the site. There should be ways for you to swap the failed component with a hot standby spare component. If you have built or bought a subsystem that has built-in redundancy and automatically fails over to the secondary component when the primary component fails, it should communicate and report this fact to you immediately so that you are aware of it right away. You may have been in the middle of an exposure, so you need to understand the failure in real time so that you can evaluate the impact, if any, on the data you are acquiring. It is important that the system fails gracefully so that you can assess the problem and successfully recover the system if necessary, or shut it down reliably if required.

As discussed in the section "Additional SSCs for Remotely Located Observatories," having some sort of remotely operated power switch to drop power and restore power might be all that is needed to restore operations. Some components simply need you to recycle power to the component to do a hardware reset and restore proper operation. Several solutions are available to accomplish this, and information is available on the Internet about these devices. It may be prudent to provide several of these components instead of relying on one overall power switch to shut down and start up your whole system. Being able to recycle power on just your imaging train components, or on your mount, or just on your communications equipment may be just the ticket to provide for easy recovery from an intermittent failure. This could definitely save your some grief in having to travel to your observatory to cycle the power.

Facing Reality—"Things Sometimes Break"

When things do go wrong with your observatory, as they always will, especially during the initial startup and commissioning of your AIS equipment, it is always nice to know that you have made your problem resolution easier by adhering to certain design philosophies. These philosophies make it much easier to troubleshoot and fix any problems that may occur. In addition, certain design features can contribute greatly to ensuring that failure modes are of the soft type where only certain parts of your AIS fail and can be recovered. This mitigates having a catastrophic failure that drops your whole observatory. As discussed in the sections "Additional SSCs for Remotely Located Observatories" and "Design for Operations—The Importance of "Not Being There "," DIDD goes a long way toward allowing for graceful failures that leave you enough of the system available to perform a controlled shutdown and perhaps a restart that will restore full operations.

You should consider providing spare parts for any long lead time equipment that may fail and could mean the difference between being down one session or down for a few weeks. Having spares on hand gives you the peace of mind that if necessary, you can be back up in running in a few hours so that you do not interrupt your observing program. This is especially relevant if your observing program includes any transient objects that are observed immediately after their discovery.

The initial commissioning of your observatory can be a trying endeavor, but if you stick to it and solve all the problems as they come up, you will be rewarded with a smooth running system that you will be proud to own and operate. You will learn a lot about your equipment during the startup phase of your observatory; don't short change yourself by creating work-arounds or bypassing the issues. It is much better to resolve the issue as it occurs rather than ignoring it, because it will eventually become a thorn in your side and you will have to fix it anyway. There's no time like the present to fix any issue that arises.

The Perfect Remote Observatory

No matter how much work you put into your remote observatory, there is no such thing as a perfect system. However, you can, with some perseverance and patience, create a system that is very reliable and will provide years of excellent service.

Further Reading

Arditti D (2007) Setting-up a small observatory. Springer
Berry R, Burnell J (2005) The handbook of astronomical image processing. Willmann-Bell
Buchheim R (2007) The sky is your laboratory. Springer
Chromey FR (2010) To measure the sky. Cambridge University Press

Covington MA (1999) Astrophotography for the amateur. Cambridge University Press
Dragesco J (1995) High resolution astrophotography. Cambridge University Press
Dymock R (2010) Asteroids and dwarf planets and how to observe them. Springer
Hubbell GR (2012) Scientific astrophotography: how amateurs can generate and use professional imaging data, Springer
Smith GH, Ceragioli R, Berry R (2012) Telescopes, eyepieces and astrographs. Willmann-Bell

Websites

Cherry Mountain Observatory, www.cherrymountainobservatory.com
iTelescope, www.itelescope.net
LightBuckets, www.lightbuckets.com
New Mexico Skies, www.nmskies.com
Sierra Remote Observatories, www.sierra-remote.com
Sierra Stars Observatory Network (SSON), www.sierrastars.com
Slooh, main.slooh.com
University of Iowa Robotic Observatory (Rigel), astro.physics.uiowa.edu/rigel
University of Arizona Mt. Lemmon SkyCenter, Skycenter.arizona.edu
Warrumbungle Observatory, www.tenbyobservatory.com
Winer Observatory, www.winer.org
ADM Accessories, http://admaccessories.com/
Apogee Imaging Systems, http://www.ccd.com/
Astro Tech, https://www.astronomytechnologies.com/
Astro Hutech/BORG, http://www.sciencecenter.net/hutech/
Astro-Physics, Inc., http://www.astro-physics.com/
Atik Cameras, http://www.atik-usa.com/
Celestron, http://www.celestron.com/
Daystar Filters, http://www.daystarfilters.com/
Denkmeier Optical, Inc., http://www.deepskybinoviewer.com/
Explora-Dome (Poly Tank), http://www.exploradome.us/
Explore Scientific LLC, http://www.explorescientific.com/
Finger Lakes Instrumentation LLC, http://www.fli-cam.com/
Hotech Corp, http://www.hotechusa.com/
Howie Glatter's Laser Collimators, http://www.collimator.com/
Innovations Foresight LLC, http://www.innovationsforesight.com/
IOptron, http://www.ioptron.com/
Lunt Solar Systems LLC, http://www.luntsolarsystems.com/
Meade Instruments, http://www.meade.com/
Moonlite Telescope Accessories, http://www.focuser.com/
Peterson Engineering Corp, http://www.petersonengineering.com/sky/index.htm
Planewave Instruments, http://www.planewaveinstruments.com/
QHYCCD (Astro Factors/Deep Space Products), http://www.astrofactors.com
QSI, http://qsimaging.com/
Questar, http://www.questar-corp.com/
Santa Barbara Instrument Group, http://www.sbig.com/
Shelyak Instruments, http://pagesperso-orange.fr/shelyak/en/index.html
Sky Watcher, http://www.skywatcher.com/
Software Bisque, http://www.bisque.com/
Starlight Instruments LLC, http://www.starlightinstruments.com/
Starlight Xpress Ltd., http://www.sxccd.com/

Stellarvue, http://www.stellarvue.com/
Tele Vue Optics, http://www.televue.com/
Texas Nautical/Takahashi America, http://www.takahashiamerica.com/
Vernonscope, http://www.vernonscope.com/
Vixen Optics, http://www.vixenoptics.com/
William Optics, http://www.williamoptics.com/

Chapter 5

Operating Your
Remotely Controlled
Observatory

The Need for Procedures—Real-Time Operations

When you own and operate complex systems such as your remotely controlled observatory and Astronomical Imaging System (AIS), it is important that you operate it consistently and with great care to avoid possibly damaging it or, at the least, obtaining poor data results. One solution is to use procedures in operating your equipment. Defined, repeatable procedures help ensure your actions produces consistent results so that you do not inadvertently introduce errors into the process as a result of an "ad hoc" style of operating.

This is not to say that once you are trained in the procedure and you have become thoroughly familiar with it that you cannot operate your system from memory. At that point, you use the procedure as a checklist to verify that you have completed the required actions. When you operate equipment in a real-time fashion, locally or remotely, there are time factors that may affect your results. A good example is when you are taking your calibration frames prior to your imaging session. Specifically, when you acquire "twilight flats," you are relying on the brightness of the twilight sky immediately after sunset, and you only have a fixed number of minutes (typically 15–20 min) to complete your acquisition of the flats.

This time element means that you need to have a thoroughly tested and reliable method of operating your equipment to acquire that data in the time allotted at a specific time. Prerequisites you need to be complete in preparation include setting up your AIS, getting your data acquisition software running, and making sure your camera system is fully operational. Also, you need to set up the storage location for your data for the session. All these actions must be completed by a specific time so

© Springer International Publishing Switzerland 2015
G.R. Hubbell et al., *Remote Observatories for Amateur Astronomers*, The Patrick Moore Practical Astronomy Series, DOI 10.1007/978-3-319-21906-6_5

that you can successfully take your twilight flats. This is just one example of many cases where you will be dealing with a time element in your operations, and it is important for you to have a tried-and-true, tested procedure that will provide consistent, high-quality results.

Situational Awareness—"Knowing What's Going On"

Another benefit that procedures provide, whether you are using them from memory or have them in hand for step-by-step use, is "situational awareness." Procedures allow you to understand exactly where you are in the process of operating your systems and help you keep track of every parameter and element of your equipment. This, in turn, enables you to immediately identify and respond to problems with your equipment or any other unplanned event that may occur with your AIS. Procedures also help you document configurations and other needed parameters that affect the operation of your observatory and may also be hard to remember from day to day because of the changes you make periodically in equipment, software, project focus, etc.

Procedures save you the time and hassle of trying to figure out what went wrong with a dataset or why your system did not perform as expected. Based on your previous experience with successful operations, you have a baseline to compare subsequent operations and quickly identify and resolve any issues.

Changing Configurations

When you decide to change configurations, the most efficient approach is to change only those items that contribute to the desired outcome. This is necessary because you probably have already tested and used the existing equipment configuration, so you do not want to introduce any "bugs" into the part of the system that you know already performs and works to your satisfaction. When introducing changes, one way to guard against unnecessary or inadvertent changes is to do regression testing. In this test procedure, you go back and re-test all the items that are part of your system to ensure that you have not introduced any problems. (See also the section "Getting Down to Details—The Observing Program Design Basis" for more on regression testing.)

It is good to step back and analyze your proposed changes to identify any issues you may possibly introduce and create a plan to mitigate those issues. If you are successful in making your change without introducing an error, then this will build confidence in your system and you know that you can rely on it at a moment's notice.

Operational Procedures

The following operational procedures are by no means a complete list of all the procedures that you may require. Also, you may not need all the procedures listed here, depending on how you plan to operate your system and based on your Observing Program Design Basis (OPDB) document. Use these procedures as they are, or consider them as a starting point for your own, custom procedures tailored to your AIS and remote observatory systems.

It is important to note that the chapter references in the procedures listed below denote chapters in the book *Scientific Astrophotography: How Amateurs Can Generate and Use Professional Imaging Data*. You will see a section in each procedure called "SA Book Chapter References," and specific references to that book in the "Notes" section of each procedure. The following is a list of each of the example procedures:

1. Image Scaling Calculation Procedure
2. AIS Equipment Setup Procedure
3. Imaging Train Configuration Procedure
4. Astrograph Collimation Procedure
5. Mount Polar Alignment Procedure
6. AIS All-Sky Alignment Procedure
7. Flat Frame Acquisition Procedure
8. Dark Frame Acquisition Procedure
9. Bias Frame Acquisition Procedure
10. Light Frame Acquisition Procedure
11. Astrograph Focusing Procedure
12. Guide-scope Setup Procedure

1. Image Scaling Calculation Procedure

Description: This task uses the specifics of the object type as input to determine the best image scaling information to acquire the best data for that object.

Objective: To calculate the image scale for the given object.

SA Book Chapter References: Chapter 5, Sections 5.3 and 5.4. Chapter 8, Section 8.2.

AIS Equipment Used: Astrograph, charge-coupled device (CCD) camera, and component equipment manuals.

Task Checklist

Step	Task	Notes
[] 1	**Use** the astrograph and camera manuals to **Record** the following information: A. Camera Pixel Size, μm B. Astrograph Focal Length, mm C. Camera, Number of Pixels x and y axis	
[] 2	**Record** the Program Object to Observe	Refer to SA Book Chapter 2 to help determine your observing program. Refer to Chapters 3–6, as required
[] 3	**Determine** the desired field of view (FOV) and pixel scale for the object recorded in Step 2	
[] 4	**Determine** the initial imaging train configuration desired	
[] 5	**Use** the data from Steps 1 and 3 to **Calculate** the image scale, pixel scale, and FOV for the imaging train configuration specified in Step 4	Refer to SA Book Chapter 5, Sections 5.3 and 5.4, as required
[] 6	**Ensure** that the back focus distance and other distances specified for each optical component in the owner's manual are used in configuring the imaging train	Refer to the owner's manual for the applicable optical components as described in SA Book Chapters 7 and 8, as required
[] 7	**Add** and/or **Exchange** components in the imaging train to obtain the desired FOV and pixel scale as specified in Step 3	Refer to SA Book Chapter 8, Section 8.2, as required
[] 8	**Repeat** Steps 5 and 6 as necessary to obtain the required imaging train configuration	

Acceptance Criteria:
The pixel scale and FOV are sufficient to image the Program Object specified in Step 2. The expected data acquired will provide the required accuracy in terms of the astrometric and photometric performance for the desired measurement as required by the observing program.

2. AIS Equipment Setup Procedure

Description: This task steps through the process of assembling the individual components into a complete AIS ready for startup.

Objective: To assemble the AIS to the point of startup.

SA Book Chapter References: Chapter 7, Chapter 8, and Chapter 10.

AIS Equipment Used: Mount, astrograph, CCD camera, and support components, and user manuals as needed.

Task Checklist

Step	Task	Notes
[ ] 1	**Inventory** all the desired equipment that makes up the desired AIS configuration	
[ ] 2	**Assemble** the tripod/pier and mount subassembly (including counterweights). **Place** the mount in the desired location, and **Point** the Right Ascension (RA) axis toward Polaris	Keep the mount as low as possible to minimize vibration levels
[ ] 3	**Level** the mount according to the user manual	This is important to ensure a repeatable polar alignment without having to adjust the Altitude axis on the mount for every session
[ ] 4	**Place** the astrograph on the mount's adapter plate as required. **Tighten** the adapter plate fasteners to ensure the astrograph does not move	
[ ] 5	**Perform Procedure 3** to determine the components for the imaging train. **Assemble** the imaging train components as specified, including CCD camera, filter wheel, coupling adapters, and other components as necessary	While referring to SA Book Chapter 7, perform **Procedure 3** as required
[ ] 6	**Install** the imaging train subassembly on the astrograph	
[ ] 7	**Install** any secondary components on the astrograph (finders, dew shield)	
[ ] 8	**Install and Connect** any electronic and electrical cabling required by the imaging train components and mount. **Ensure** that the power supplies are fully charged and/or available for use	Installation includes running the cables as necessary to minimize the chance of snagging a cable
[ ] 9	**Balance** the astrograph with all its component subassemblies in the RA and Declination (DEC) axes	Establish a small imbalance in RA and DEC to help minimize the periodic error and backlash effects while imaging. Refer to SA Book Chapter 6, Section 6.2.2, for information on periodic error
[ ] 10	**Start up** all AIS control software and **Ensure** that the electronic equipment is operational prior to use	

Acceptance Criteria:

The AIS components are integrated and tightly coupled to their adjacent components. There is *no* play between components. All the cabling is secure and free from snagging when moved about the full range of motion in the RA and DEC axes.

3. Imaging Train Configuration Procedure

Description: This task steps through the process of assembling the components that make up the specific imaging train for a given observing program object.

Objective: To assemble the imaging train for installation into the AIS.

SA Book Chapter References: Chapter 7 and Chapter 8.

AIS Equipment Used: CCD camera, filter wheel, and other major and support optical and mechanical components used to create the complete imaging train.

Task Checklist

Step	Task	Notes
[ ] 1	**IF** not already done, **THEN Perform Procedure 1**	**Procedure 1** is used to calculate the image scale required for the program object observed and guides you in selecting the required components
[ ] 2	**Select** all the components that will be used to assemble the imaging train. These components must meet the requirements determined by performing **Procedure 1**	
[ ] 3	**Inspect** each major component's interface (threads/barrels/locking screws) for damage and cleanliness. **Correct** any issues before assembly	**CAUTION:** The mechanical connections between components must be free of damage and debris before assembly or permanent damage may occur
[ ] 4	**Place** each major component and support mechanical and optical component on a clean surface in its proper position as determined by **Procedure 1**	
[ ] 5	**Assemble** the imaging train as laid out in Step 4, carefully tightening each component as necessary to avoid damaging the interfacing threads/barrels/locking screws	Refer to SA Book Chapters 7 and 8, as required
[ ] 6	**Verify** that the measurements for back focus distance and the placement for the optical components meet the requirements of the optical component in the owner's manual	Refer to SA Book Chapters 7 and 8, as required
[ ] 7	**Inspect** the final assembly for completeness, tightness, and orientation. **Correct** any issues before installation of the imaging train	The orientation of the imaging train on the AIS should be considered before mounting it on the AIS to ensure that the imaging train components do not interfere with the slewing of the astrograph
[ ] 8	**Place** the imaging train in a safe location until ready to install on the AIS	

Acceptance Criteria:

The imaging train is configured according to the requirements determined in the performance of Procedure 1. The imaging train is assembled as required to perform the observations as desired and specified in Procedure 1. The imaging train is rigid and does not exhibit any flexure or motion when applying slight pressure to the components. The CCD camera is square to the mounting point as desired, where the imaging train meets the astrograph focuser.

4. Astrograph Collimation Procedure

Description: This task steps through the various processes used for collimating the astrograph.

Objective: To collimate the astrograph.

SA Book Chapter References: Chapter 7 and Chapter 8.

AIS Equipment Used: Astrograph, Cheshire eyepiece, medium focal length (FL) eyepiece, imaging train, and/or collimation software used in conjunction with CCD images acquired for this objective.

Task Checklist

Step	Task	Notes
[ ] 1	**NOTE:** The collimation procedure is generally only regularly applicable to a reflector type astrograph. Refractors do not normally require re-collimation on a regular basis. Refer to the owner's manual to determine the need to collimate a refractor **IF** not already done, **THEN Perform Procedure 2. IF** you are doing a visual collimation using a collimation mask, **THEN Substitute** a medium FL eyepiece for the imaging train in Step 6 of Procedure 2. **IF** you are using a CCD camera to perform the collimation, **THEN Go To** Step 6	**Procedure 2** is a pre-requisite for performing the visual astrograph collimation because you MUST have an operational AIS before collimation
[ ] 2	**NOTE:** This step should be performed indoors in order to easily see the internal components of the astrograph **Point** the astrograph at a bright, white surface. Using a Cheshire eyepiece in place of the medium FL eyepiece specified in Step 1, **Identify** the internal concentric components of your astrograph according to the astrograph owner's manual. **Perform** an initial collimation using the Cheshire eyepiece	Refer to the astrograph owner's manual, as required, for the specific steps and adjustments necessary for the proper initial and final collimation of your astrograph

(continued)

(continued)

Step	Task	Notes
[ ] 3	**Insert** a medium FL eyepiece into your astrograph's focuser. **Slew** your astrograph to a bright star and **Adjust** the focus so that the star is out of focus. When the focus is inside or outside the optimal focus, the star appears as a donut-shaped object when using an astrograph with a central obstruction, or as a disk when using a refractor	Observing the donut-shaped star image, or disk (as applicable) within the medium FL eyepiece allows you to determine the amount of offset in the collimation and guide you in the fine adjustment of the secondary of your astrograph
[ ] 4	**Adjust** the astrograph's collimation controls as necessary to bring the star image to a concentric image shape according to the astrograph owner's manual to achieve optimal collimation	
[ ] 6	**CAUTION:** It is necessary to have a CCD camera in your imaging train that is capable of high image acquisition rates, and/or sub-frame imaging to complete the adjustment in a timely manner **Perform Procedures 1, 2, and 3** before performing the collimation using a CCD camera	It is necessary to have available an operational AIS, including an imaging train that can be used for collimation purposes to perform **Procedures 1, 2, and 3**
[ ] 7	**Ensure** that the applicable CCD camera imaging software is running and available for use. Before use, **Functionally Test** the imaging software to ensure its proper operation	
[ ] 8	**Adjust** the imaging software to acquire images at a rate necessary to provide adequate feedback on the adjustments made during the collimation process. **IF** a proper update frequency cannot be obtained, **THEN** use the sub-frame features of the imaging software to increase the image acquisition rate	Usually a rate of at least 1 frame per second (fps) is necessary to receive adequate feedback on your adjustments. Slower rates should be avoided. Refer to the imaging program owner's manual as required
[ ] 9	**Adjust** the focus so that the star is out of focus. When the focus is inside or outside the optimal focus, the star appears as a donut-shaped object or disk (as applicable)	Observing the star image using the imaging software allows you to determine the amount of offset in the collimation and guides you in the fine adjustment of the secondary mirror, or primary refractive element of your astrograph
[ ] 10	**Adjust** the astrograph's collimation controls as necessary to bring the star image to a concentric image shape on the imaging software's display. **Adjust** the image according to the astrograph owner's manual to achieve optimal collimation	Refer to the imaging program owner's manual, as required. Look for any special features the imaging program may have to aid in the collimation of the astrograph

Acceptance Criteria:

The image provided by the medium FL eyepiece or the imaging software depicts a properly collimated astrograph as described and depicted in the astrograph owner's manual. Other depictions of a properly collimated astrograph are also available on the Internet.

5. Mount Polar Alignment Procedure

Description: This task steps through the process of performing an accurate polar alignment of the AIS mount using the CCD Drift Method.

Objective: To polar align the AIS mount using the CCD Drift Method.

SA Book Chapter References: Chapter 6, Chapter 8, and Chapter 11. Refer to http://canburytech.net/DriftAlign/index.html for a thorough explanation of the drift alignment method.

AIS Equipment Used: Complete AIS system, including imaging train.

Task Checklist

Step	Task	Notes
[ ] 1	**IF** not already done, **THEN Perform Procedures 1, 2, and 3** before performing the polar alignment using a CCD camera. **Perform** a collimation as necessary using **Procedure 4**	A fully functional AIS imaging train is necessary to perform an accurate polar alignment
	NOTE: There are several different methods for performing a high-precision polar alignment but this procedure uses the Drift Method. This procedure can also be performed visually using a cross-hair eyepiece instead of the CCD camera, although it is easier to monitor progress with the CCD camera. Ensure that the FOV of the camera is at least 20 by 20 arc-minutes for the initial exposure	Refer to SA Book Chapter 6, Section 6.3. Refer to the mount's owner's manual or the mount's ASCOM driver manual for proper operation of the mount controller
[ ] 2	**NOTE:** The following steps are used to determine the direction and amount of drift necessary for determining the adjustment needed on the **Azimuth Axis** of the mount	Generally, you can expect excellent image data if you expose your light frames at one half the total exposure time (TET) used to polar align your mount with this method
	Slew the astrograph to point to a relatively bright star near the Meridian and the Celestial Equator. **Enable** the Sidereal Tracking Rate on the mount controller if necessary	

(continued)

(continued)

Step	Task	Notes
[ ] 3	**Adjust** the position of the mount to center a star in the CCD camera's FOV. **Enable** the cross-hair feature of the imaging software. **Set** the imaging software to **Acquire** images at a suitable rate to monitor the position of the star	Refer to the imaging software owner's manual. A rate of at least 1 fps is suggested for image acquisition during this portion of the procedure
[ ] 4	**Disable** the Sidereal Tracking Rate on the mount controller and **Monitor** the direction of star drifts on the display	
[ ] 5	**Rotate** the imaging train as necessary to **Align** the long edge of the CCD camera and cross-hair with the drift axis. This is the RA axis	The image edge corresponding to the direction the star is drifting toward is the **WEST**
[ ] 6	**Enable** the Sidereal Tracking Rate on the mount controller	
[ ] 7	**Re-Position** the star to the **eastern** edge of the image FOV using the RA and DEC buttons on the mount controller as necessary	
[ ] 8	**Center** the star on the DEC axis of the image. **Note** the direction the star moves on the image display when moving the mount in the north direction (Pressing the North Controller button)	The image edge corresponding to the direction toward which the star is drifting is the **SOUTH**
[ ] 9	**Set** the imaging software to acquire a 60-s exposure. **Set** the Slew Rate for the RA axis on the mount controller to 2× Sidereal	The purpose of the long exposure is to be able to acquire a star trail that enables you to graphically see the difference between the polar axis and the mount's RA axis
[ ] 10	**Monitor** the timer for the exposure. After the following elapsed times, **Perform** the actions specified:	
[ ] 11	**Start** the 60-s exposure	
[ ] 12	**After 10 s (elapsed)**—**Disable** the Sidereal Rate on the mount controller	
[ ] 13	**After 35 s (elapsed)**—**Press and Hold** the mount controller W button to **Slew** the mount to the **WEST** at the set rate of 2× Sidereal	
[ ] 14	**After 60 s (elapsed)**—At the completion of the exposure, **Release** the mount controller W button and **Enable** the Sidereal Rate	

(continued)

(continued)

Step	Task	Notes
[ ] 15	**Examine** the image to verify that you have a horizontal, long, V-shaped star trail with a bright star image at the beginning of one of the legs in the V. **Determine** the direction of DEC drift **IF** the V leg without the bright star is toward the **SOUTH** edge, **THEN** the star drifted **SOUTH**, otherwise it drifted **NORTH**	
[ ] 16	**(A) Adjust** the mount's Azimuth axis control a small amount to move the mount toward the **EAST** if the star drifted **NORTH** **(B) Adjust** the mount's Azimuth axis control a small amount to move the mount toward the **WEST** if the star drifted **SOUTH**	**IF** performing this correction when located in the Southern Hemisphere, **THEN Reverse** the Azimuth axis directions (EAST should be WEST and vice versa)
[ ] 17	**Re-Position** the star to the eastern edge of the image FOV and to the center on the DEC axis	
[ ] 18	**Repeat** Steps 9–17 to close the gap between the legs in the V-shaped star trail. **Adjust** the TET in Steps 9 and 11 from 60 s up to 600 s as necessary to obtain the polar alignment accuracy required. **IF** the exposure time is greater than or equal to the calculated value, **THEN** a portion of the V-shaped trails will not be within the FOV of the CCD camera. This will not affect the adjustment; simply perform Step 16 as specified	**Adjust** the elapsed time (ET) in seconds for Steps 13 and 14 as follows to expose equal star trail lengths on each of the legs of the V: Step 13 ET sec $= 10 + ((TET - 10)/2)$ sec Step 14 ET sec $= TET$ sec
[ ] 19	**Slew** the astrograph to point to a relatively bright star to the EAST (90° Azimuth) or WEST (270° Azimuth), 20–30° above the horizon. **Enable** the Sidereal Tracking Rate on the mount controller if necessary	The following steps are used to determine the direction and amount of drift necessary for determining the adjustment needed on the **Altitude axis** of the mount The eastern horizon is preferred because the stars are rising and the effects of refraction are lessened while doing the adjustment
[ ] 20	**Re-Position** the star to the **EAST** edge of the image FOV using the RA and DEC buttons on the mount controller as necessary	
[ ] 21	**Center** the star on the DEC axis of the image	

(continued)

(continued)

Step	Task	Notes
[ ] 22	**Set** the imaging software to acquire a 60-s exposure. **Set** the Slew Rate for the RA axis on the mount controller to 2× Sidereal	The purpose of the long exposure is to be able to acquire a star trail that enables you to graphically see the difference between the polar axis and the mount's RA axis
[ ] 23	**Monitor** the countdown timer for the exposure. After the following elapsed times, **Perform** the actions specified:	
[ ] 24	**Start** the 60-s exposure	
[ ] 25	**After 10 s (elapsed)—Disable** the Sidereal Rate on the mount controller	
[ ] 26	**After 35 s (elapsed)—Press and Hold** the mount controller W button to **Slew** the mount to the **WEST** at the set rate of 2× Sidereal	
[ ] 27	**After 60 s (elapsed)**—At the completion of the exposure, **Release** the mount controller W button and **Enable** the Sidereal Rate	
[ ] 28	**Examine** the image to verify that you have a horizontal, long, V-shaped star trail with a bright star image at the beginning of one of the legs in the V. **Determine** the direction of DEC drift. **IF** the V leg without the bright star is toward the **SOUTH** edge, **THEN** the star drifted **SOUTH**, otherwise it drifted **NORTH**	
[ ] 29	(A) **IF** the star drifted NORTH, *AND* the star pointed to is on the eastern horizon, **THEN Adjust** the mount's Altitude axis control a small amount to **LOWER** the mount toward the ground (B) **IF** the star drifted SOUTH, *AND* the star pointed to is on the eastern horizon, **THEN Adjust** the mount's Altitude axis control a small amount to **RAISE** the mount toward the sky (C) **IF** the star drifted NORTH, *AND* the star pointed to is on the western horizon, **THEN Adjust** the mount's Altitude axis control a small amount to **RAISE** the mount toward the sky (D) **IF** the star drifted SOUTH, *AND* the star pointed to is on the western horizon, **THEN Adjust** the mount's Altitude axis control a small amount to **LOWER** the mount toward the ground	**IF** performing this correction when located in the Southern Hemisphere, **THEN Reverse** the Altitude axis directions (**NORTH** should be **SOUTH** and vice versa) Depending on the case, Perform A, B, C, or D
[ ] 30	**Re-Position** the star to the EAST edge of the image FOV and to the center on the DEC axis	

(continued)

(continued)

Step	Task	Notes
[ ] 31	**Repeat** Steps 22–30 to close the gap between the legs in the V-shaped star trail. **Adjust** the TET in Steps 22 and 24 from 60 s up to 600 s as necessary to obtain the polar alignment accuracy required. **IF** the exposure time is greater than or equal to the calculated values, **THEN** a portion of the V-shaped trails will not be within the FOV of the CCD camera. This will not affect the adjustment; just perform Step 29 as specified	**Adjust** the ET in seconds for Steps 26 and 27 as follows to expose equal star trail lengths on each of the legs of the V: Step ET sec = $10 + ((TET - 10)/2)$ Step ET sec = TET

Acceptance Criteria:

At the end of the total exposure time, the imaged star trails overlap each other, indicated by no widening of the trail at either end. This indication is applicable only for the longest exposure time required by the program object you plan to image. The longest exposure time used in doing the polar alignment should be two times the longest expected light frame exposure time.

6. AIS All-Sky Alignment Procedure

Description: This task steps through the process of performing an all-sky alignment (synchronization) to ensure the mount points to the RA and DEC coordinates commanded by the mount controller. The all-sky alignment compensates for bias, orthogonal, and cone errors in the astrograph-mount interface.

Objective: To synchronize the AIS pointing to the true sky coordinates.

SA Book Chapter References: Chapter 6, Section 6.2.

AIS Equipment Used: AIS, mount, mount hand controller, and mount's hand controller owner's manual, planetarium program or database for reference coordinates, and/or astrometric software.

Task Checklist

Step	Task	Notes
[ ] 1	**IF** not already done, **THEN Perform Procedures 1, 2, and 3** before performing the all-sky alignment using a CCD camera. **Perform** a collimation as necessary using **Procedure 4**	A fully functional AIS imaging train is necessary to perform an accurate all-sky alignment

(continued)

(continued)

Step	Task	Notes
[ ] 2	**NOTE:** A three-star alignment corrects for bias, orthogonal, and cone errors in the astrograph-mount interface. There are two ways to do the all-sky alignment— using the mount controller automation, and performing a manual three-star all-sky alignment. This procedure provides the steps for both **IF** a visual alignment using a cross-hair eyepiece and the mount's hand controller is required, **THEN Proceed** to the next step. Otherwise, for an accurate CCD all-sky alignment, **Proceed** to Step 8	Most hand controllers for go-to mounts provide a built-in procedure for doing a two- or three-star alignment. This is usually performed using the astrograph visually with an eyepiece
[ ] 3	Using the mount's hand controller, **Select** the three-star alignment procedure	Refer to the mount's hand controller owner's manual as required
[ ] 4	Using the mount's hand controller, **Select** and **Slew** to the first star of the alignment procedure	
[ ] 5	Using the astrograph finder and cross-hair eyepiece, **Center** the star in the FOV, and when prompted by the hand controller, **Synchronize** to the coordinates of the star selected	
[ ] 6	**Repeat** Steps 4 and 5 for the next two stars	
[ ] 7	**Follow** the hand controller's prompts to complete the three-star alignment procedure; **THEN** go to Step 15	Refer to the mount's hand controller manual for instructions on performing a three-star alignment
[ ] 8	**Ensure** that the mount's ASCOM or other proprietary controller software and the planetarium or other program used to select star coordinates are up and running	Refer to the mount's ASCOM and other driver information in the mount's owner's manual
[ ] 9	**Connect** the planetarium program or other star coordinate program to the mount. **Functionally Test** the Un-park, Park, and Slew commands using the program interface	
[ ] 10	**Start** the imaging train's imaging software. **Set** the imaging software to **Acquire** images at a suitable rate to monitor the position of the star	Refer to the imaging software owner's manual A rate of at least 1 fps is suggested for image acquisition during this portion of the procedure
[ ] 11	**Un-Park** the mount and **Slew** to the first of three stars to perform the all-sky alignment	If using a planetarium program to control your mount, this is usually accomplished by pointing and clicking on the target star and selecting the SLEW TO command

(continued)

(continued)

Step	Task	Notes
[ ] 12	**Enable** the cross-hair feature of the imaging software. Using the mount's controller, **Center** the star image in the cross-hair display	Refer to the image acquisition and planetarium program owner's manual for information on how to use the Plate Solve functions of the program(s) if available
		As an alternative to centering the image of the star, use the astrometric software to **Perform** a Plate Solve and **Transfer** the coordinates to the planetarium program (in the next step). In this case, at least a 30- to 60-s exposure is required to acquire enough stars in the image to perform the Plate Solve
[ ] 13	**Select SYNC** in the planetarium program to synchronize the star's position and transfer the coordinates to the planetarium program	
[ ] 14	**Repeat** Steps 11–13 for two more stars	
[ ] 15	**Select and Slew** to a bright target and **Verify** that the target is within the FOV of the CCD camera or eyepiece as necessary	

Acceptance Criteria:

The mount's pointing is successfully synchronized using three stars with the mount's hand controller or the planetarium software. The mount can successfully point to a selected target by slewing the astrograph to the required coordinates by verifying that the selected target is near the center of the FOV of the eyepiece or CCD camera as required.

7. Flat Frame Acquisition Procedure

Description: This task steps through the process of acquiring the flat frames used to calibrate the raw imaging data acquired in the light frames.

Objective: To acquire the calibration flat frames.

SA Book Chapter References: Chapter 13, Sections 13.5 and 13.6.

AIS Equipment Used: AIS, imaging train, flat frame light source, light diffuser.

Task Checklist

Step	Task	Notes
[] 1	**NOTE:** **IF** not already done, **THEN Perform Procedures 1, 2, and 3. Perform** a collimation as necessary using **Procedure 4**	This procedure must be done with the complete imaging train attached to the astrograph. The flat frame includes data to compensate the light frame for systematic and random errors, including bias, vignetting, filter non-uniformity, and other defects. There are two sources of light for taking the flat frame, a light panel and the twilight sky. This procedure is applicable to both sources of light A fully functional AIS imaging train is necessary when acquiring flat frames for image calibration
[] 2	**Adjust** the focuser to a position close to the expected focus position that will be used for acquiring the light frames	Record the focus position when making light frame exposures and refer to these positions when adjusting the focuser for this procedure
[] 3	**Set** the CCD camera temperature to the desired value depending on the season. **Enable** the CCD camera temperature controller software to maintain this temperature	Refer to the CCD camera temperature control section in the imaging software's owner's manual as required
[] 4	**Record** the temperature value in the observing log	
[] 5	**Un-Park and Slew** the astrograph position to zenith if acquiring twilight sky-flats, or to point to the flat light source, as desired	
[] 6	**Place** a light diffuser over the front of the astrograph as desired	
[] 7	**CAUTION:** If the CCD camera has a mechanical shutter, an exposure time of at least 5 s is necessary to mitigate the effects of shadowing (gradients) when exposing the CCD at too low a shutter speed **Set** the exposure time in the imaging software to **Acquire** images to provide a MAX ADU equal to 80–90 % of full-well depth, or 50,000–55,000 ADUs (205–230 for an 8-bit CCD)	If taken during twilight, exposures in the range of 3–10 s should be expected
[] 8	**Define and Create** the directory (as necessary) where the flat frames are to be stored during this imaging session	
[] 9	**Start** the image capture process and **Acquire** at least 10 images at the desired ADU value	

(continued)

Step	Task	Notes
[ ] 10	**Open and Examine** (as necessary) one of the flat frames for quality	
[ ] 11	After acquiring all the images desired, **Park** the mount	

Acceptance Criteria:

A minimum of 10 flat frames were acquired, each with a MAX ADU count of at least 80 % of full-well depth and checked for quality.

8. Dark Frame Acquisition Procedure

Description: This task steps through the process of acquiring the dark frames used to calibrate the raw imaging data acquired in the light frames.

Objective: To acquire the calibration dark frames.

SA Book Chapter References: Chapter 13, Sections 13.5 and 13.6.

AIS Equipment Used: Astrograph coupled to the fully configured imaging train.

Task Checklist

Step	Task	Notes
[ ] 1	**OPTIONAL: IF** not already done, **THEN Perform Procedures 1, 2, and 3. Perform** a collimation as necessary using **Procedure 4**	**NOTE 1:** The dark frame includes data to compensate the light frame for systematic and random errors, including bias and thermal dark current electrons **NOTE 2:** This procedure can be performed without mounting the imaging train on the astrograph and indoors as long as the TEC temperature setting is set to a specific value that will be used when acquiring light frames A fully functional AIS imaging train is necessary when acquiring dark frames for image calibration
[ ] 2	**OPTIONAL: Adjust** the focuser to the in-focus position close to the expected focus position used for acquiring the light frames	
[ ] 3	**Set** the CCD camera temperature to the desired value depending on the season. **Enable** the CCD camera temperature controller software to maintain this temperature	Refer to the CCD camera temperature control section in the imaging software's owner's manual as required
[ ] 4	**Record** the temperature value in the observing log	

(continued)

(continued)

Step	Task	Notes
[ ] 5	**Set** the CCD camera imaging software to keep the shutter closed (as necessary) and **Cover** the astrograph with the dust cover to block all light from entering the CCD camera	
[ ] 6	**Set** the exposure time to 60 s in the imaging software to **Acquire** the dark frame images	
[ ] 7	**Define and Create** the directory (as necessary) where the dark frames are to be stored during this imaging session	
[ ] 8	**Start** the image capture process and **Acquire** at least ten images	
[ ] 9	**Open and Examine** (as necessary) one of the dark frames for quality	

Acceptance Criteria:

A total of at least ten dark frames were acquired and checked for quality.

9. Bias Frame Acquisition Procedure

Description: This task steps through the process of acquiring the bias frames used to calibrate the raw imaging data acquired in the light frames.

Objective: To acquire the calibration bias frames.

SA Book Chapter References: Chapter 13, Sections 13.5 and 13.6.

AIS Equipment Used: Astrograph coupled to the fully configured imaging train.

Task Checklist

Step	Task	Notes
[ ] 1	**OPTIONAL: IF** not already done, **THEN Perform Procedures 1, 2, and 3. Perform** a collimation as necessary using **Procedure 4**	**NOTE 1:** The bias frame includes data to compensate the light frame for systematic errors **NOTE 2:** This procedure can be performed without mounting the imaging train on the astrograph and indoors as long as the TEC temperature setting is set to a specific value that will be used when acquiring light frames A fully functional AIS imaging train is necessary when acquiring bias frames for image calibration
[ ] 2	**OPTIONAL: Adjust** the focuser to the in-focus position close to the expected focus position used for acquiring the light frames	

(continued)

(continued)

Step	Task	Notes
[ ] 3	**Set** the CCD camera temperature to the desired value depending on the season. **Enable** the CCD camera temperature controller software to maintain this temperature	Refer to the CCD camera temperature control section in the imaging software's owner's manual as required
[ ] 4	**Record** the temperature value in the observing log	
[ ] 5	**Set** the CCD camera imaging software to keep the shutter closed (as necessary) and **Cover** the astrograph with the dust cover to block all light from entering the CCD camera	
[ ] 6	**Set** the exposure time to zero (0) seconds in the imaging software to **Acquire** the bias frame images	
[ ] 7	**Define and Create** the directory (as necessary) where the bias frames are to be stored during this imaging session	
[ ] 8	**Start** the image capture process and **Acquire** at least 30 images	
[ ] 9	**Open and Examine** (as necessary) one of the bias frames for quality	

Acceptance Criteria:

A minimum of 30 bias frames were acquired and checked for quality.

10. Light Frame Acquisition Procedure

Description: This task steps through the process of acquiring the light frames containing the raw data for the object under study.

Objective: To acquire the program object raw data.

SA Book Chapter References: Chapter 13, Sections 13.3, 13.4, and 13.7.

AIS Equipment Used: Complete AIS as configured for acquiring program object images.

Task Checklist

Step	Task	Notes
[ ] 1	**IF** not already done, **THEN Perform Procedures 1, 2, and 3. Perform** a collimation as necessary using **Procedure 4**	A fully functional AIS imaging train is necessary for this procedure
[ ] 2	**Adjust** the focuser to the in-focus position close to the expected focus position used for acquiring the light frames	

(continued)

(continued)

Step	Task	Notes
[ ] 3	**IF** available, **Set** the CCD camera temperature to the desired value depending on the season. **Enable** the CCD camera temperature controller software to maintain this temperature	Refer to the CCD camera temperature control section in the imaging software's owner's manual as required
[ ] 4	**Record** the temperature value in the observing log as required	
[ ] 5	**Define and Create** the directory (as necessary) where the light frames are to be stored during this imaging session	
[ ] 6	**Perform Procedure 11** upon initial setup and periodically during the imaging session	Checking and adjusting the focus periodically ensures that the best data are being acquired
[ ] 7	**Set** the exposure time to the desired value in the imaging software to **Acquire** the light frame images. **Set** the video gain, exposure time, and frame rate for the camera as necessary when using a webcam type CCD camera	Refer to SA Book Chapter 10 for recommendations on the exposure times for various objects
[ ] 8	**Start** the image capture process as needed according to the specific observing program object requirements	
[ ] 9	**Monitor** the data acquisition process to ensure that the light frames are of the required quality	

Acceptance Criteria:

The light frames are acquired and checked for the required quality.

11. Astrograph Focusing Procedure

Description: This task steps through the process of adjusting the astrograph focuser to achieve an accurate focus to acquire the best imaging data possible.

Objective: To adjust the astrograph focuser to achieve an accurate focus.

SA Book Chapter References: Chapter 7, Section 7.9, and Chapter 3, Section 3.3.

AIS Equipment Used: AIS, AIS focuser and applicable focusing software, focusing mask, and imaging software.

Task Checklist

Step	Task	Notes
[ ] 1	**Perform Procedures 1, 2, and 3** before performing the all-sky alignment using a CCD camera. **Perform** a collimation as necessary using Procedure 4	There are two ways to focus a star accurately using the AIS focusing subsystem: using the statistics of the star image directly (FWHM, and Peak ADU), and by using a focusing mask A fully functional AIS imaging train is necessary when focusing an image using the image acquisition software
[ ] 2	**Un-Park** the mount and **Slew** the astrograph to a bright star	
[ ] 3	**Ensure** that the applicable CCD camera imaging software is running and available for use. Before use, **Functionally Test** the imaging software to ensure its proper operation	
[ ] 4	**Enable** the cross-hair feature of the imaging software. Using the mount's controller, **Center** the star image in the cross-hair display. **Place** the focusing mask on the front of the astrograph if using that method	
[ ] 5	**Adjust** the imaging software to acquire images at a rate necessary to provide adequate feedback of the adjustments made during the focusing process. **IF** a proper update frequency cannot be obtained, **THEN** use the sub-frame features of the imaging software to increase the image acquisition rate	Usually a rate of at least 0.2 fps (5-s interval between frames) is necessary to receive adequate feedback on your adjustments. Slower rates should be avoided. Refer to the imaging program owner's manual as required
[ ] 6	**Select** the star on the display and **Monitor** the imaging programs statistics feature providing the **FWHM** and **Peak ADU** value for the star used to focus the astrograph. Alternatively, **Observe** the diffraction lines that the focusing mask creates to aid in focusing the star image	Many imaging programs provide a special dialog box showing a selected star's statistics used for focusing. Refer to the imaging program's owner's manual for this specific feature
[ ] 7	**Adjust** the astrograph focuser position to **Minimize** the FWHM value and/or **Maximize** the Peak ADU value for the star under study. Alternatively, **Adjust** the focuser to bring the diffraction lines of the focusing mask to the proper position	Refer to the focusing mask instructions to learn how the image should appear when properly focused
[ ] 8	Once the focuser is adjusted for best focus, **Lock Down** the focuser using the focuser's lock-screw	

Acceptance Criteria:

The star is properly focused as indicated by the value of the FWHM of the star being MINIMIZED, and/or the Peak ADU being MAXIMIZED. As specified in the owner's manual, the focusing mask view depicts a properly focused star image.

12. Guide-Scope Setup Procedure

Description: This task steps through the process of setting up and initial calibration of the guide-scope.

Objective: To set up and calibrate the guide-scope subsystem.

SA Book Chapter References: Chapters 3, 4, and 5; Chapter 6, Sections 6.1 and 6.3; Chapter 7, Sections 7.9, 7.11, and 7.12.

AIS Equipment Used: AIS, mount, guide-scope, and auto-guiding application software.

Task Checklist

Step	Task	Notes
[ ] 1	**IF** not already done, **THEN Assemble** the components that make up the guide-scope subsystem	The components are the guide-scope, guide camera, associated power and signal cabling, and mounting system
[ ] 2	**IF** not already done, **THEN Mount** the assembled guide-scope subsystem onto the AIS	
[ ] 3	**IF** not already done, **THEN Apply** power to the guide-scope subsystem	
[ ] 4	**IF** not already done, **THEN Start Up and Connect** to the guide camera using the auto-guiding software	There are several different sources for auto-guiding software, standalone and integrated into image acquisition software
[ ] 5	**Select** a bright star and **Slew** the astrograph to the star	
[ ] 6	**Center** and **Focus** the star's image in the guide camera using the guide-scope's focuser. **Adjust** the frame rate as necessary to obtain the best image	
[ ] 7	**Start** the auto-guiding system's software calibration routine to **Calibrate** the mount control response to the guider's inputs	The auto-guiding system calibration is dependent on the side of the Meridian to which the guide-scope is pointed because of the Meridian flip that occurs when crossing the Meridian. Refer to the auto-guiding software owner's manual for detailed instructions on calibrating the system Whenever the German equatorial mount (GEM) performs a Meridian flip, recalibration of the auto-guiding system is required
[ ] 8	**Select and Lock** onto a star to guide on using the auto-guiding software's controls	

Acceptance Criteria:
The successful completion of the calibration process leads to successfully guiding on a selected star using the auto-guiding software and guide-scope.

Suggested Equipment Lists

Your OPDB will be the source document for you to begin defining your system and equipment list. You may want to start with a list that has the bare essentials, often referred to in the aviation industry as the Minimum Equipment List (MEL). The MEL will list all those items needed to do only what is necessary and excludes any extra items that may be desired for comfort or convenience reasons. It is good to start with the MEL and identify the bare minimum for operation so you can concentrate on getting that right before you add any other items that could complicate or add work to your process that could affect performance. It is important to obtain your desired system performance using only the items on the MEL.

Another list you may want to create is the Approved Equipment List (AEL). This is similar to the MEL but has the added pedigree of providing a list of equipment that has been configured and validated as a working, performing system. The MEL becomes the AEL once you have assembled the system and operated it with the procedures you have written for it.

One other list you may consider creating is a list of support items or logistic equipment that may not be a formal part of your operating systems but is needed to support the operation of your observatory. Examples include personal equipment such as flashlights, eye protection, gloves, and other protective clothing to match your weather and environment. Also consider other items such as food and drink, and first-aid kits, medicines, etc.

Further Reading

Arditti D (2007) Setting-up a small observatory. Springer
Berry R, Burnell J (2005) The handbook of astronomical image processing. Willmann-Bell
Buchheim R (2007) The sky is your laboratory. Springer
Byrne CJ (2005) Lunar Orbiter photographic atlas of the near side of the Moon. Springer
Chromey FR (2010) To measure the sky. Cambridge
Covington MA (1999) Astrophotography for the amateur. Cambridge
Dragesco J (1995) High resolution astrophotography. Cambridge
Dymock R (2010) Asteroids and dwarf planets and how to observe them. Springer
Harrison KM (2011) Astronomical spectroscopy for amateurs. Springer
Henden AA, Kaitchuck, RH (1990) Astronomical photometry. Willmann-Bell
Howell SB (2006) Handbook of CCD astronomy. Cambridge
Hubbell GR (2012) Scientific astrophotography: how amateurs can generate and use professional imaging data. Springer
Shirao M, Wood CA (2010) The Kaguya lunar atlas. Springer
Smith GH, Ceragioli R, Berry R (2012) Telescopes, eyepieces and astrographs. Willmann-Bell
Warner BD (2006) A practical guide to lightcurve photometry and analysis. Springer
Warner BD (2010) The MPO user's guide. BDW Publishing

Websites

Cherry Mountain Observatory, www.cherrymountainobservatory.com
iTelescope, www.itelescope.net
LightBuckets, www.lightbuckets.com
New Mexico Skies, www.nmskies.com
Sierra Remote Observatories, www.sierra-remote.com
Sierra Stars Observatory Network (SSON), www.sierrastars.com
Slooh, main.slooh.com
University of Iowa Robotic Observatory (Rigel), astro.physics.uiowa.edu/rigel
University of Arizona Mt. Lemmon SkyCenter, Skycenter.arizona.edu
Warrumbungle Observatory, www.tenbyobservatory.com
Winer Observatory, www.winer.org
ADM Accessories, http://admaccessories.com/
Apogee Imaging Systems, http://www.ccd.com/
Astro Tech, https://www.astronomytechnologies.com/
Astro Hutech/BORG, http://www.sciencecenter.net/hutech/
Astro-Physics, Inc., http://www.astro-physics.com/
Atik Cameras, http://www.atik-usa.com/
Celestron, http://www.celestron.com/
Daystar Filters, http://www.daystarfilters.com/
Denkmeier Optical, Inc., http://www.deepskybinoviewer.com/
Explora-Dome (Poly Tank), http://www.exploradome.us/
Explore Scientific LLC, http://www.explorescientific.com/
Finger Lakes Instrumentation LLC, http://www.fli-cam.com/
Hotech Corp, http://www.hotechusa.com/
Howie Glatter's Laser Collimators, http://www.collimator.com/
Innovations Foresight LLC, http://www.innovationsforesight.com/
IOptron, http://www.ioptron.com/
Lunt Solar Systems LLC, http://www.luntsolarsystems.com/
Meade Instruments, http://www.meade.com/
Moonlite Telescope Accessories, http://www.focuser.com/
Peterson Engineering Corp, http://www.petersonengineering.com/sky/index.htm
Planewave Instruments, http://www.planewaveinstruments.com/
QHYCCD (Astro Factors/Deep Space Products), http://www.astrofactors.com
QSI, http://qsimaging.com/
Questar, http://www.questar-corp.com/
Santa Barbara Instrument Group, http://www.sbig.com/
Shelyak Instruments, http://pagesperso-orange.fr/shelyak/en/index.html
Sky Watcher, http://www.skywatcher.com/
Software Bisque, http://www.bisque.com/
Starlight Instruments LLC, http://www.starlightinstruments.com/
Starlight Xpress Ltd., http://www.sxccd.com/
Stellarvue, http://www.stellarvue.com/
Tele Vue Optics, http://www.televue.com/
Texas Nautical/Takahashi America, http://www.takahashiamerica.com/
Vernonscope, http://www.vernonscope.com/
Vixen Optics, http://www.vixenoptics.com/
William Optics, http://www.williamoptics.com/

Part II

Using Remote
Observatory Facilities

Chapter 6

Accessing Professional Observatories Around the Globe

Remote Observing—Pushing Down the Technology

Historically, large professional observatories are designed to support astronomy research and education programs. These facilities are expensive to build and maintain, and researchers and students usually have a short, fixed time to use them. In many cases, observing time must be scheduled months to a year or more in advance. If the scheduled time is not useable, owing to weather or technical problems, the opportunity is lost because the schedule queue can be fully booked well into the future. This supply and demand conflict is restrictive and prioritized, which can frustrate people whose research and careers are affected by not getting adequate telescope time to realize their goals.

The supply and demand problem is a result of how expensive large observatory facilities are to build, run, and maintain. Older generation telescopes and observatories built before the 1990s required a team of skilled technicians to operate. Most require at least some manual intervention to set up and control an observing run and are not capable of running in a fully automated mode.

The latest generation of telescopes and observatories, using new electronic devices controlled by sophisticated integrated computer hardware and software systems, can run in an automated mode and be controlled remotely. Even though these observatories require less manual operation and maintenance, they still require a staff of skilled technicians to oversee the systems, troubleshoot and fix hardware and software problems, clean optics, recoat mirrors, and so on. As a result, the costs are roughly the same or even greater than the older generation observatories. However, the new generation of telescopes is much lighter and

© Springer International Publishing Switzerland 2015
G.R. Hubbell et al., *Remote Observatories for Amateur Astronomers*, The Patrick Moore Practical Astronomy Series, DOI 10.1007/978-3-319-21906-6_6

capable of higher precision tracking and pointing, which is vital for automated observing runs.

The development of charge-coupled device (CCD) imaging cameras is likely the greatest advance in observatory imaging technology since the invention of photography. Photographic emulsion films used in astronomy up until the 1990s had low quantum efficiencies (QE) compared with CCD imaging technology. QE measures the percentage of photons "hitting" a film or chip that are "captured" and counted. The highest QE of the most sensitive film emulsions used in astronomy was about 3 %. A technique called hyper-sensitization can push the QE of some film up to about 10 % at best.

In contrast, CCD chips used in astronomical imaging achieve QEs of 50 to 97+ percent. CCD technology offers an additional advantage over film. Film suffers from reciprocity failure, which, in effect, causes the QE to drop the longer the film is exposed. On the other hand, CCD chips register photons linearly until saturated and thus do not suffer from reciprocity failure.

Finally, CCD technology is digital and does not require chemical processing as film does. You can take CCD images, and store and view them within seconds after an exposure. Because of their high QE and other advantages of CCD chips, modest size telescopes (20- to 40-in. (0.5- to 1.0-m) diameter) with CCD cameras can view targets with magnitudes as faint as the Mount Palomar 200-in. (5-m) telescope could using photographic plates for imaging.

During the twentieth century, professional observatories housed ever bigger reflecting telescopes, superseding the massive refractors of the nineteenth century, the largest of which is the 40-in. (1-m) telescope still housed in the Yerkes Observatory in Williams Bay, Wisconsin. New technology in glass materials and manufacturing, together with new technology for coating mirror surfaces with aluminum, enabled the fabrication of mirrors much larger than the lenses of refractors. The Hale 200-in. (5.1-m) telescope on Mount Palomar was the first large telescope to use low-expansion Pyrex glass and remained the largest telescope in the world for more than three decades. Its mirror weighs 29,000 lb (13,200 k). The mount is a horseshoe yoke equatorial mount made from steel. The entire assembly is massive and heavy. It was built like a battleship.

Starting in the late 1980s, advances in computers, materials science, CCD camera technology, and advanced electronic systems fostered development of a new generation of high-technology astronomical telescopes. Not only could they be manufactured in much larger sizes, but smaller ones were built lighter and could operate semi-autonomously. The age of the automated robotic observatory was born. CCD camera technology was the most revolutionary, disruptive technology of all.

Some professional observatories have retrofitted their older telescopes to incorporate the new technologies. Telescopes built before the 1980s were operated manually by people who set them up and "babysat" them all night while someone else manually guided the telescope on objects. Upgrading this type of telescope to the precision needed for automated operation is difficult and expensive. Often it is simply not possible.

Today's modern professional observatories are integrated electronic systems. The operation of all the observatory components—telescope, CCD camera, filter wheel, dome, and so on—is controlled by sophisticated software programs and high-precision electronic components. In effect, the entire observatory becomes an automated robotic system.

Automating an observatory is a game changer. Integrating the systems under a single, centralized control system reduces the cost of operating the observatory. It eliminates the need for one or more people to operate the observatory onsite during an observing run. Onsite personnel can also be unnecessary for observatories controlled remotely but not running in an automated queuing mode.

Remotely accessible professional observatory telescopes set up for use by amateur and professional astronomers on demand only became available in the twenty-first century. Many of these observatories have strict rules about who uses their systems, when you can use them, how often you can use them, and so on. The exception is the growing number of commercial operations offering remote access for a fee. Many of these telescopes are relatively small. However, a couple of these companies now offer advanced-technology telescopes as large as 32 in. (0.8 m) in diameter. These are big telescopes, by amateur astronomer standards, situated in observatory systems that cost hundreds of thousands of dollars and require expert technicians to oversee and maintain them. In the future, these entrepreneurial companies might offer even larger telescopes for their customers' to use (Fig. 6.1).

Today, the number of 20- to 40-in. (0.5- to 1.0-m) diameter telescopes owned by individuals and institutions is increasing. Before the 1990s, telescopes of this size range were primarily housed in professional observatories operated by colleges and institutions. The 200-in. (5.1-m) Hale telescope on Mount Palomar was the largest regularly operating telescope until 1990 and started using CCD-technology imaging systems in the early 1980s. At the time, a 40-in. (1-m) telescope was considered a moderately big telescope.

How to Access the Professional Scopes

There are two methods for accessing and using remote observatory telescopes: direct control and automated queued scheduling.

Direct Remote Control

Some observatory systems are set up to enable you, the user, to control remote telescopes directly. In these cases, you or your group gains direct access to control a telescope from a remote location. You control operation using a software interface, which often is a web browser-based Internet connection. The user interface usually has some kind of control panel that gives you feedback on the status of the

Fig. 6.1 SSON 0.8-meter Mount Lemmon SkyCenter Remote Observatory (Credit Adam Block)

telescope, such as where the telescope is pointing, what filter is active, weather conditions, and so on. It also has controls to point the telescope to an object or set of coordinates, select filters, set exposure times, and so on (Fig. 6.2).

Controlling a remote telescope directly is the only option if you insist on hands-on telescope control. Real-time control gives you a true sense of "telepresence" while operating the telescope. However, during the entire block of time you control the telescope directly, it is unavailable to others. This limits how many users can access the telescope in an observing session and when users can image objects during an observing session.

Direct remote control works well for projects where you want to take multiple images of an object contiguously as is the case when running large sets of images for luminance, red, green, and blue (LRGB) color composite images. Direct control helps mitigate time lost selecting new objects, slewing to new locations, refocusing,

Fig. 6.2 Several iTelescope direct remote control telescopes in roll-off roof observatory (Credit iTelescope)

and so on. Another advantage is the ability to image what you want, when you want, on the fly.

There are, however, downsides to the direct control method. One of these is you must be awake and ready to operate the telescope during your allotted time. This is inconvenient if what you want to study is best observed when you would normally be asleep. As mentioned earlier, because only one user (or group of users) can control a telescope at any given time, no one else can image during that time. This can be a problem in cases where an interesting object is only observable for a short period during an observing run. For example, comets are typically at their brightest when they are closer to the Sun and, therefore, low in the sky near the start or end of astronomical twilight. If the window of opportunity to image a comet is about an hour and someone has control of the telescope for 40 min of that time, it leaves little time for anyone else to image the comet.

Perhaps the most challenging type of imaging using direct control is taking a series of images over long periods of time with set intervals. For example, many professional and amateur astronomers do asteroid search and follow-up projects that require imaging a field of view (FOV) at a specific set of coordinates to detect and measure the position of an asteroid. Their discovery and follow-up technique is to take a set of three or four images spaced 10–15 min apart and then repeat the process about an hour later. This gives them the type of datasets the Minor Planet

Table 6.1 Advantages and disadvantages of direct remote control

Advantages	Disadvantages
You have hands-on direct control of the telescope with the feeling of telepresence	Only one person or group can operate the telescope during the time the telescope is under remote control
You point where you want while in command	Taking long image series over many hours is impractical and tedious
You can efficiently take many back-to-back images in a short period of time	Depending on where users are located, the time of the observing window may be inconvenient
This is a good method for observing rare transient events quickly	

Center (MPC) requires for proper astrometry measurements to determine orbital elements for the object.

Many amateur astronomers around the world do asteroid light curve projects to determine the rotation and, with sufficient data collected over many months, the shape of the asteroids. This work requires taking an image of the target asteroid every few minutes or so for up to several hours. Running a project like this directly could occupy a telescope most of an observing run and would be a tedious task for the user. Table 6.1 summarizes the advantages and disadvantages of direct remote control.

Automated Queued Control

Many professional observatories run observing sessions using a queue-based observatory scheduling control system. You submit an observing request using an online entry form to specify an object or set of coordinates, filters, exposure times, number of exposures to take, time spacing between exposures, and so on. All the observing requests are then submitted to a master scheduling software program. This program creates a master schedule incorporating all the acceptable schedule requests. It schedules each image observation to run as close to the optimum time as possible for the specific observing run. This method is the most efficient to acquire the most images in an observing run. It enables the acquisition of many images around the same time with a minimum of wasted time in between. It also enables dozens of users' schedule requests to run in an observing session by interleaving exposures within tight timeframes when needed.

Remote professional observatories available to the general public are a scarce, valuable resource. Automated queue-based control systems allow dozens of people to use a telescope during the same observing run. They handle many more users' needs in a night than a direct remote-control system can. For example, images for a large introductory college astronomy class laboratory project can easily be run in a single night without anyone competing for the time (Fig. 6.3).

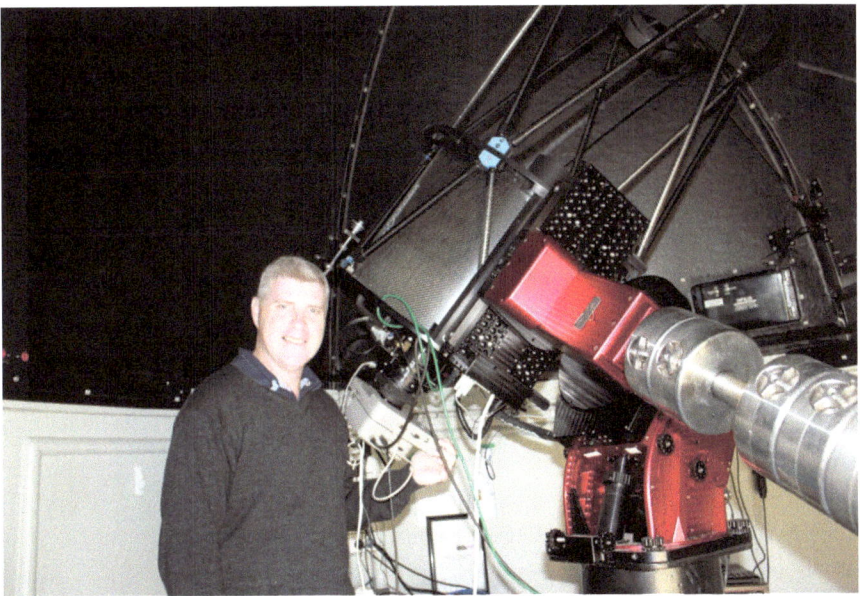

Fig. 6.3 SSON 0.5-meter telescope in the Warrumbungle Observatory in Coonabarabran, Australia (Credit Peter Starr)

Queue-based schedules are perfect for projects requiring precise intervals over long periods of time. Professional sky survey and near Earth object (NEO) search programs use this type of scheduling control. They can take hundreds of images in an observing run without the need for someone to control the telescopes.

Automated control of observing schedules eliminates the tedium of running long precise schedules manually under remote control. Also, observations can run while the users are doing something else or sleeping while helping eliminate mistakes and saving time.

In addition, these systems typically allow you to schedule observations for specific times and dates. This is an important consideration for projects such as determining the exact minima of a binary star eclipse or imaging an extrasolar-planet transit that occurs at specific predicted times. In these cases, you want to take a series of images that runs for a period of time before and after the event to catch the entire process. Because the timing and dates are known to a certain precision beforehand, you can schedule your observing runs accordingly. For projects such as asteroid light curves, where you want to collect lots of image data over weeks or months, you can create schedules to run on future dates without worrying about what you will be doing at the time.

Table 6.2 Advantages and disadvantages of automated queue control

Advantages	Disadvantages
You set the schedules for the images you want to receive, and the work is done for you	You do not have the flexibility of directly controlling a telescope in real time
There is no need to actively operate a remote telescope	You cannot react quickly to short transient events such as gamma ray bursts
You can take long series of images with precise intervals	
Specific times and dates can be set for schedules to run without you having to be available to do the work	
More telescope time is available for more people	

In summary, automated queue-controlled observatory systems are the best option if you are more interested in acquiring image data than in operating a telescope. Table 6.2 lists the advantages and disadvantages of direct remote control.

What's Best for Your Project?

The comparison of direct control and automated queue-controlled telescopes shows that each has advantages and disadvantages. Which system is best for your project depends on what you want to accomplish.

If you are new to astronomy or simply want to explore the sky gathering your own image data, using a direct remote-control system is a good place to start. You can learn a lot by imaging interesting objects, especially the more spectacular objects in the sky, and processing the data to produce your own unique images. You also experience the joy of operating a sophisticated professional telescope at your command. It is an empowering experience for anyone new to astronomy.

However, use of direct remote-control systems is not limited to beginners. As you learn how to use these systems more effectively and efficiently, they become excellent tools for generating images using various filters to create impressive color composite images after proper image processing. Some astronomers use these systems to image unusual and uncommon events, such as gamma ray bursts and other transient events.

For many serious professional and amateur astronomers, the benefits of automated queue control outweigh those of the direct control systems. Typically, they are more interested in getting high-quality imaging automatically for their projects than operating a telescope directly. If you compare the advantages and disadvantages of these two methods, the reasons for the preference are apparent. As stated earlier, controlling a telescope directly to take exposures at precise intervals over a long period of time is tedious and demands close attention to get the timing right. Some of the types of projects best suited for queue-based imaging include:

- Asteroid discovery and follow-up
- Comet dynamics over extended periods
- Asteroid light curves
- Eclipsing binary star and extrasolar-planet transits
- Continual regular monitoring of galaxies to find new supernovae
- Regular photometry measurements of large numbers of variable stars

More Factors to Consider

In addition to choosing whether to use a direct control or queue-based system for getting your image data, there are other issues to consider for your project:

- How faint is the object you want to image?
- How big an FOV do you need for your project?
- Do you require special filters for your project?
- Is your object of interest located in the Northern or Southern Hemisphere sky?

In general, the diameter of the telescope mirror or objective lens and the QE (sensitivity to registering photons) of the CCD camera chip determine what your exposure times need to be to achieve a desired signal-to-noise ratio (SNR). SNR is a number that expresses how much stronger the light of your target object is (the signal) than the background noise in the image. For photometry work, where you want to measure the brightness of a star or other object, you typically want to achieve an SNR of 100 or better for measurements with a precision of 0.01 magnitude or better. For astrometry work, where you are only trying to determine the position of the object, you can get by with a SNR of 10 or less to get precise measurements.

A larger-diameter telescope using a camera with the same QE as a smaller-diameter telescope in similar seeing conditions produces images with a higher SNR for the same exposure times. Therefore, the larger-diameter telescope also shows fainter objects in the same exposure times.

The size of the CCD chip in a camera determines the size of the FOV of acquired images in telescopes with the same focal length. In addition, with the same size CCD chip, long focal length telescopes have a smaller FOV than short focal length telescopes. For example, most astronomy CCD camera manufacturers offer cameras with the popular KAF-16803 CCD chip. This chip contains 4096 by 4096 9-μm pixels, which creates a chip size of 1.5 by 1.5 in. (36.8- by 36.8 mm). A Takahashi FS-152 apochromatic refractor has a 6-in. (152-mm) lens and a 47.9-in. (1216-mm) focal length (F/8). The pixel resolution with this combination is 1.5 arc-seconds per pixel. The resulting FOV is 104 by 104 arc-minutes or 1.73 degrees. This is a very wide FOV—nearly 3.5 times greater than the width of a full Moon. For comparison, the 32-in. (0.81-m) F/7 Sierra Stars Observatory Network (SSON) Mount Lemmon SkyCenter Schulman telescope has a much longer focal length of 223 in. (5670 mm). This telescope has a CCD camera with the same KAF-16803 chip but with a pixel resolution of 0.33 arc-seconds per pixel. The FOV with this combination—22.5 by 22.5 arc-minutes or 0.38 degrees—is considerably smaller than with the same chip used on the Takahashi telescope.

Also keep in mind that although a longer focal length gives a smaller FOV for a given size CCD chip, the image scale increases at the same time. In other words, longer focal lengths make objects in an image appear bigger. This effect is sometimes referred to as *magnification*. Larger image scales enable you to see detail in small galaxies, planetary nebula, and other objects that might not appear in small image scales.

The filter wheels on remote telescopes hold a limited number of filters—most hold between five and eight filters. For photometry projects, many people want to use standard Johnson-Cousins ultraviolet-blue-visual-red-infrared (UBVRI) filters. These filters have coatings that only pass light in certain defined bandwidths so measurements by different telescopes can be compared in a standardized way. People interested in doing aesthetic imaging projects often use a set of LRGB filters to produce color composites. The L filter (luminance) is a clear filter with a coating to block infrared light. Some telescope systems also have narrowband filters such as H-alpha, OIII (oxygen), and SII (sulfur) for special imaging projects.

No single ground-based observatory can image the entire sky. If the object you want to image is located near the south celestial pole, you will not be able to image it with a Northern Hemisphere observatory. Fortunately, some of the commercially available remote observatory networks have telescopes located in both hemispheres.

Cost Considerations

In general, companies that offer remote astronomy imaging services charge prices that increase with the size of the telescope used. This makes sense because larger telescopes are more expensive to buy and maintain. You should choose the size of the telescope to match the needs of your observing program. If you decide to image brighter objects or you want to image very wide FOVs, a refractor or smaller (less than 20-in. (0.5 m)) Cassegrain telescope might fit your needs well. To image and measure faint objects dimmer than 19th magnitude, telescopes of 20-in. (0.5 m) and larger diameter are the best choice.

Most of these companies charge at an hourly rate in the form of credits, typically in US dollars. You create an account and buy credits that are deducted from your account after you use the telescopes or receive delivery of the images, which are usually stored on a server where you can download them.

Some companies offer a membership/subscription while others use a more straightforward direct "pay for image data" method. Also, some companies offer a discount rate system based on the Moon phase and other parameters that can be confusing for some people to figure out.

In 2007, SSON was the first company to offer the most straightforward and least expensive method for charging its users. SSON charges you for the actual exposure times of the images that you schedule and receive. There is no additional overhead cost. Also, SSON was the first company to implement a policy of reimbursing credits for images that are of poor quality, regardless of the reason. Recently, other companies have adopted this model, which has benefited all remote imaging users.

How much you pay for your imaging data depends on the policies of the company you use and the telescope you choose for your imaging program. So how do you know what is the most cost-effective telescope to use? The short answer is—it depends.

If your program goal is to get the best image data using various filters to create an esthetically pleasing color composite image of a galaxy or nebula, the more data you collect and combine, the more detail and contrast you'll see in the final product. However, at a certain point, adding more images produces only a slight increase in the image quality. If you use a large telescope that costs $100 per hour for exposure time, and 90 min of total exposure time produces a final image only slightly better than 60 min total time, is the final result worth the $50 difference? This is a subjective decision that only you can make.

For scientific photometry and astrometry programs, determining the cost of images to achieve the results you want is more straightforward. Your objective is either to measure the precise brightness (photometry) or position (astrometry) of an object. The precision of a photometry measurement or the ability to measure the position of a very faint object (such as an asteroid) is determined by the SNR of the object in the image. The greater the signal (brightness) of the object measured compared with the background noise, the more precise your measurements are. The SNR you achieve for each telescope depends on several factors, including the diameter of the telescope, QE (sensitivity) of the CCD camera, sky brightness, seeing, and other factors. Describing how to calculate the SNR of telescope/CCD camera systems is beyond the scope of this book. However, you can find SNR calculators by searching on the Internet.

To determine which remote telescope/camera system best meets the needs of your program, do some "what if" SNR calculations before scheduling telescope time. The SNR you want to achieve depends on the minimum precision that is acceptable for your measurement goal. For example, an SNR of 100 equates to a precision of 0.01 magnitude for your measurements. If you want to achieve this precision as a minimum, then all your image data should have an SNR of 100 or higher. On the other hand, you can achieve high-precision astrometry measurements of objects with an SNR of 10 or less.

So how does SNR relate to the cost of using remote telescopes for your program? It depends on how faint an object is and what precision you want to achieve. For example, let's say the objective of your program is to measure the light curve to determine the minimum magnitude of an eclipsing binary star system you expect to dim to approximately 16th magnitude, and you want to measure the event to a precision of 0.01 magnitude using a V filter for photometry and get the measurement as close to the exact minimum as possible.

For this example, let's compare two telescopes in the SSON system with different sizes and rates per hour to determine which is most appropriate to get the best results. The two telescopes are a 14.5-in. F/14 (0.35-m) Cassegrain telescope and a 24-in. F/10 (0.61-m) Cassegrain telescope. The QE and the FOV of the CCD cameras on the two telescopes are comparable, and they each have a V filter with the same Johnson-Cousins photometry specifications. The smaller telescope rate is $50/h for exposure time, and the larger telescope rate is $100/h.

To keep the comparison equal, let's assume that both sites image without any moonlight, have the same sky brightness (21.5) and seeing (2 arc-seconds), and observe the same object at the same air mass (1.2). Under these conditions, the telescopes achieve the following SNR for 120-s images:

- 14.5-in. (0.35-m) telescope = ~108 SNR
- 24-in. (0.61-m) telescope = ~189 SNR

In this case, the smaller telescope meets the minimum SNR requirement at half the cost of the larger telescope. Therefore, the smaller telescope is the most cost-effective telescope to use for this program.

However, let's say that you need more data points on the light curve to achieve a smoother and more precise light curve and to measure the timing of the minimum to a precision of about 1 min. Because the exposure times are 120 s, the number of data points in your light curve measurements and resolution of the timing don't meet your needs. The solution is to use the larger telescope to achieve the desired SNR with shorter exposure times. If you set the exposure times for each telescope under the same conditions to 40 s you get the following SNR calculations:

- 14.5-in. (0.35 m) telescope = ~63 SNR
- 24-in. (0.61 m) telescope = ~109 SNR

The SNR of the smaller telescope now falls below the desired precision you want while the larger telescope meets the requirement with room to spare. As you can see, to save yourself time and money, it's good practice to do some "what if" calculations before you schedule images or operate a remote telescope.

Further Reading

Arditti D (2007) Setting-up a small observatory. Springer
Berry R, Burnell J (2005) The handbook of astronomical image processing. Willmann-Bell
Buchheim R (2007) The sky is your laboratory. Springer
Byrne CJ (2005) Lunar Orbiter photographic atlas of the near side of the Moon. Springer
Chromey FR (2010) To measure the sky. Cambridge
Covington MA (1999) Astrophotography for the amateur. Cambridge
Dragesco J (1995) High resolution astrophotography. Cambridge
Dymock R (2010) Asteroids and dwarf planets and how to observe them. Springer
Harrison KM (2011) Astronomical spectroscopy for amateurs. Springer
Henden AA, Kaitchuck, RH (1990) Astronomical photometry. Willmann-Bell
Howell SB (2006) Handbook of CCD astronomy. Cambridge
Hubbell GR (2012) Scientific astrophotography: how amateurs can generate and use professional imaging data, Springer
Shirao M, Wood CA (2010) The Kaguya lunar atlas. Springer
Smith GH, Ceragioli R, Berry R (2012) Telescopes, eyepieces and astrographs. Willmann-Bell
Warner BD (2006) A practical guide to lightcurve photometry and analysis. Springer
Warner BD (2010) The MPO user's guide. BDW Publishing

Websites

Cherry Mountain Observatory, www.cherrymountainobservatory.com
iTelescope, www.itelescope.net
LightBuckets, www.lightbuckets.com
New Mexico Skies, www.nmskies.com
Sierra Remote Observatories, www.sierra-remote.com
Sierra Stars Observatory Network (SSON), www.sierrastars.com
Slooh, main.slooh.com
University of Iowa Robotic Observatory (Rigel), astro.physics.uiowa.edu/rigel
University of Arizona Mount Lemmon SkyCenter, Skycenter.arizona.edu
Warrumbungle Observatory, www.tenbyobservatory.com
Winer Observatory, www.winer.org
ADM Accessories, http://admaccessories.com/
Apogee Imaging Systems, http://www.ccd.com/
Astro Tech, https://www.astronomytechnologies.com/
Astro Hutech/BORG, http://www.sciencecenter.net/hutech/
Astro-Physics, Inc., http://www.astro-physics.com/
Atik Cameras, http://www.atik-usa.com/
Celestron, http://www.celestron.com/
Daystar Filters, http://www.daystarfilters.com/
Denkmeier Optical, Inc., http://www.deepskybinoviewer.com/
Explora-Dome (Poly Tank), http://www.exploradome.us/
Explore Scientific LLC, http://www.explorescientific.com/
Finger Lakes Instrumentation LLC, http://www.fli-cam.com/
Hotech Corp, http://www.hotechusa.com/
Howie Glatter's Laser Collimators, http://www.collimator.com/
Innovations Foresight LLC, http://www.innovationsforesight.com/
IOptron, http://www.ioptron.com/
Lunt Solar Systems LLC, http://www.luntsolarsystems.com/
Meade Instruments, http://www.meade.com/
Moonlite Telescope Accessories, http://www.focuser.com/
Peterson Engineering Corp, http://www.petersonengineering.com/sky/index.htm
Planewave Instruments, http://www.planewaveinstruments.com/
QHYCCD (Astro Factors/Deep Space Products), http://www.astrofactors.com
QSI, http://qsimaging.com/
Questar, http://www.questar-corp.com/
Santa Barbara Instrument Group, http://www.sbig.com/
Shelyak Instruments, http://pagesperso-orange.fr/shelyak/en/index.html
Sky Watcher, http://www.skywatcher.com/
Software Bisque, http://www.bisque.com/
Starlight Instruments LLC, http://www.starlightinstruments.com/
Starlight Xpress Ltd., http://www.sxccd.com/
Stellarvue, http://www.stellarvue.com/
Tele Vue Optics, http://www.televue.com/
Texas Nautical/Takahashi America, http://www.takahashiamerica.com/
Vernonscope, http://www.vernonscope.com/
Vixen Optics, http://www.vixenoptics.com/
William Optics, http://www.williamoptics.com/

Chapter 7

Matching Your Observing Program to the Available Professional Equipment

Chapter 6 described the advantages and disadvantages of the two basic methods of using remotely located observatories (direct and queue-based control). This information can help you design an observing program that best meets the needs of the observations you have in mind. Before starting out on your program, you should evaluate what you want to accomplish and develop a realistic plan that meets your goals. Some people like to spend time up front to outline a plan for a project in great detail before starting out. Others prefer to jump right in with a general idea and figure things out using a kind of "seat of the pants" approach. There is no one "right way" to plan and implement your observing program. Either method can work for you. However, if you start out blindly with no plan, you will waste time and money.

Having a plan—either in a detailed written outline or in your head—gives you a roadmap to get to your ultimate goal. Whichever method you use, there are fundamental questions you need to answer to create a plan that makes sense. Here are some of the most important ones:

- Is your goal to create an esthetically pleasing image or is it to get precise scientific measurements from the image data?
- Can you complete your program goals in a single night or a few observing sessions, or is it a long-term program that might require weeks or months to accomplish?
- Do you want to control a remote telescope directly for the immediacy and control it provides or are you mainly concerned with getting high-quality data without the need to control a telescope directly?

© Springer International Publishing Switzerland 2015
G.R. Hubbell et al., *Remote Observatories for Amateur Astronomers*, The Patrick Moore Practical Astronomy Series, DOI 10.1007/978-3-319-21906-6_7

- Does your project require access to telescopes located in the Northern Hemisphere, Southern Hemisphere, or both?
- What size telescope is appropriate to reach the signal-to-noise ratio (SNR) you want to achieve in a reasonable amount of time?
- What type of filters will you use?
- Is your object of interest well placed in the sky for the time of year you plan to image it?
- How much will it cost to get the image data you need to achieve your project goals?

The answers to these questions enable you to choose the remote observatory telescopes that best match your needs.

Matching the Remote Observatory to Your OPDB

To illustrate how to use these questions to determine which remote observatories match your needs, this section describes two types of programs. Each example answers the questions to determine the kind of observatory the "program manager" should choose.

Monitor the Magnitude and Behavior of Faint Comets

This advanced program aims to image comets as faint as 16th magnitude and take photometric measurements using standard Johnson-Cousins filters.

What Is the Project Goal?
The goal is to determine the change in brightness of comets and to monitor them for changes in behavior, such as an outburst or the development of a tail.

What Is the Estimated Time to Complete the Program?
This is an ongoing program with an unknown duration. The number of comets monitored will vary from 10 to 20 or more.

Direct Control or Queued Sessions?
Direct remote control is impractical because of the number of comets imaged and because they are often located low in the sky in the early evenings and early mornings with a short window of opportunity to get high-precision image data. Automated queued sessions enable the collection of the image data with minimal effort.

Does the Project Require a Northern Hemisphere Observatory, Southern Hemisphere Observatory, or Both?
Both. Comets appear in both hemispheres and sometimes switch hemispheres after perihelion. If you want to monitor such comets, you need to identify two observatories, one in the Northern Hemisphere and the other in the Southern Hemisphere, that offer similar capabilities.

What Size Telescope Is Appropriate to Reach the SNR You Want to Achieve in a Reasonable Amount of Time?

Because some of the comets will be as dim as 16th magnitude and the goal is to acquire image data with an SNR of at least 100 for high-precision photometry, the program requires a telescope with an aperture of at least 0.5 m to keep exposure times reasonable.

What Type of Filters Will You Use?

The program requires standard Johnson-Cousins R (red) and V (visual) for photometric measurements. In addition, a clear filter (unfiltered) is required to obtain brighter images to monitor comets behavior and morphology.

Is Your Object of Interest Well Placed in the Sky for the Time of Year You Plan to Image It?

This is a big variable. Near perihelion, comets are close to the Sun and appear low in the sky at the beginning or end of astronomical twilight with a short window available for imaging. Other comets farther out from the Sun may appear anywhere in the sky and thus provide a better opportunity for more precise measurements.

How Much Will It Cost to Get the Image Data You Need to Achieve Your Project Goals?

The local astronomy club allocates $100 per month for the project in which several members participate.

What Other Considerations Are Important for the Project?

The telescope must be capable of tracking a comet as it moves against the star field.

Take Images of M51 to Create a Color Composite Image

This is a first-time program for an amateur astronomer new to image processing.

What Is the Project Goal?

The goal is to get high-quality images to combine and create a beautiful color composite image.

What Is the Estimated Time to Complete the Program?

This is a first attempt using a remote observatory system. The objective is to get all the required image data in a single session.

Direct Control or Queued Sessions?

You want to control the telescope directly to get the hands-on experience of driving a telescope. The experience of controlling the telescope is as important to you as getting image data.

Does the Project Require a Northern Hemisphere Observatory, Southern Hemisphere Observatory, or Both?

The object of interest, M51, requires the use of a Northern Hemisphere observatory.

What Size Telescope Is Appropriate to Reach the SNR You Want to Achieve in a Reasonable Amount of Time?

The goal is to create a high-quality color image of M51 with a reasonable size image scale to see detail in the galaxy. Instead of worrying about what SNR to achieve as you would for scientific image data, you strive to get total exposure times long enough to create a high-contrast result. Telescopes ranging from a 6-in. (15-cm) refractor to a 16-in. (0.4-m) reflector would be appropriate for your needs.

What Type of Filters Will You Use?

You want to create a luminance, red, green, and blue (LRGB) color-composite image. Therefore, a telescope with a set of red (R), green (G), blue (B) and Luminance (L) (or clear) filters is required.

Is Your Object of Interest Well Placed in the Sky for the Time of Year You Plan to Image It?

M51 rises highest in the sky at night during the spring months. This is the best time to image it.

How Much Will It Cost to Get the Image Data You Need to Achieve Your Program Goals?

You have a small budget to try out and experiment with getting your images from a remote observatory. The best option is to use the least expensive telescope that meets your needs. After gaining experience and confidence you might opt to use bigger, more expensive telescopes for more ambitious imaging projects.

Meeting Your Observing Goals with the Equipment Available

Most professional remote observatory facilities have modern high-quality telescopes, charge-coupled device (CCD) cameras, and filters. However, the size of available telescopes varies from small refractors to large Cassegrain telescopes with mirrors approaching 1-m in diameter (and likely larger in years to come). Some telescopes have cameras with higher quantum efficiency (QE) CCD chips than others. Each telescope also has a limited set of filters that meet different needs. This burgeoning assortment of remote observatory instruments opens up many new opportunities for amateur astronomers. It also can be confusing when trying to determine what telescope to use for your program.

Small apochromatic refractors offer fast optics and very wide fields of view. They are also typically the least expensive remote telescopes to use. If you are inexperienced and just starting to experiment with using remote observatories, this might be a good option for you. Also, these wide-field refractors can image very large extended objects and star fields that you cannot capture with larger telescopes and smaller fields of view. A good example of a program for a wide-field refractor would be imaging a bright comet with a tail that extends for several degrees (Fig. 7.1).

If you want to image the faintest objects possible or get the best resolution of small objects, you need to use the larger-size reflector telescopes available. For example,

Fig. 7.1 Image of Comet Lovejoy taken with a refractor telescope

before the year 2000, amateur astronomers with 14-in. (0.35-m) to 16-in. (0.4-m) telescopes had a good chance to discover a new asteroid in their search programs. Now, several professional dedicated near-Earth asteroid (NEA) observatories search large parts of the sky every clear dark night. They now discover the majority of Main-Belt asteroids in their search for NEAs. The reason is that the professional asteroid search programs use 32-in. (0.8-m) to 80-in. (2.0-m) and larger telescopes. Some of the largest telescopes can detect and discover asteroids fainter than 23rd magnitude. The days of amateurs readily discovering new asteroids brighter than 18th magnitude now appear to be over. Today, a 24-in. (0.6-m) or larger telescope capable of imaging asteroids that are 20th magnitude or fainter is needed to compete with the professional surveys. Fortunately, remote telescopes of this size are available for you to use if you decide that you want to try discovering a new asteroid (Fig. 7.2).

In January 2013, the Sierra Stars Observatory Network (SSON) brought online the first commercial remote spectroscopy system. It was a low-resolution transmission grating spectrograph (TGS) using a 50 mm by 50 mm 600-line/mm transmission grating that fit inside the filter wheel of the 14.5-in. (0.37-m) University of Iowa Rigel telescope, which is located in southern Arizona and was part of SSON. (The Rigel telescope was replaced with a 20-in. (0.5-m) telescope in September 2015, which has a high-resolution fiber-feed spectrograph). Two TGSs were used—one without a slit and one with a 3-mm wide slit to block light except for the intended object to be imaged. Later, the same type of TGS system was added to the SSON 24-in. (0.8-m) telescope in California (Fig. 7.3).

Fig. 7.2 Catalina Sky Survey 60-in. telescope

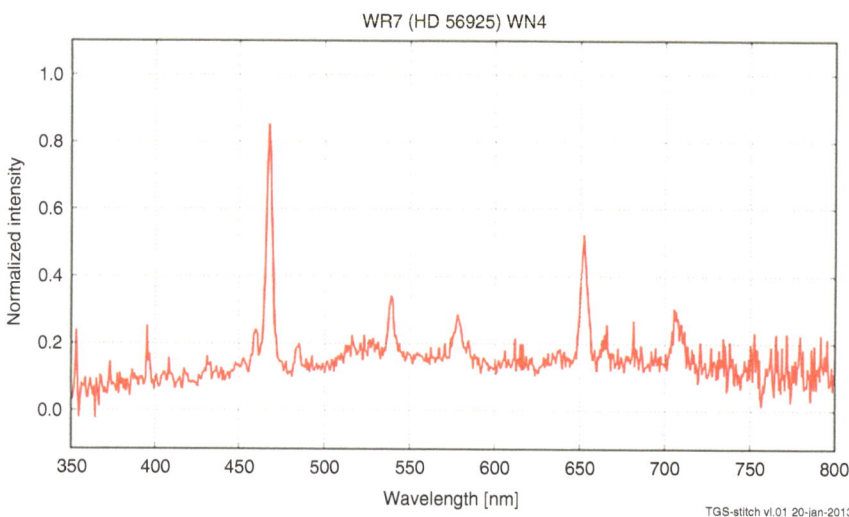

Fig. 7.3 Spectrum of WR7 (Wolf-Rayet WN2 Star) taken with the Rigel TGS (Credit Dr. Robert Mutel, Iowa State University)

Spectroscopy opens up exciting opportunities for teaching astronomy and for independent research. In the future, remote observatories may offer dedicated high-resolution spectrographs for users, which would enable more cutting-edge research opportunities.

An Introduction to Planning Your Observations

After you determine your program goals and which telescopes are most appropriate to use, you need to plan how to get the best image data from the remote observatory. If you decide to use a telescope that you will control directly, you should establish a routine that optimizes the time you spend setting up and running the telescope before you start. If you use a queue-based telescope, you have plenty of time to plan what you want to image and set all the appropriate settings before you create your image schedules.

Whichever method you use, it's a good idea to write your plan down on paper or with your favorite word processing program. You might consider keeping a detailed log of the settings you use, information about the images you take, the time you spend running the telescope (for direct control), your running costs, and so on. Everyone takes a different approach to planning. Even so, there are some specific questions you can answer to help ensure you have a solid plan that works for your observing session and your program goals.

How Long Should Your Exposure Times Be?

For a photometry or astrometry program, you determine your exposure times based on the SNR you want to achieve. If you are doing an esthetic image, your exposure time is determined by how much contrast you want to achieve to pick up faint details.

For science measurements, you can determine what the SNR will be for specific exposure times using one of the online astronomical CCD SNR calculators. The SNR calculations require you to input several variables that affect the result, including the diameter of the telescope, QE of the CCD camera, filter used, pixel size, estimated sky brightness, estimated seeing in arc-seconds, and exposure time.

For esthetic imaging projects, you generally are not concerned about the SNR, but instead want to get the best contrast and detail possible. Often that means taking exposure times as long as is practical and affordable. However, queue-based scheduling systems that are unguided typically limit exposure times to 300 s (5 min). This limit is necessary to optimize the system to get as many images as possible without the added time and complexity of using an auto-guider for longer exposures. However, this exposure limit isn't a problem because you can stack as many images as you want to create longer total exposure times. It also has the added bonus that it enables you to take long cumulative exposures of bright objects without overexposing the image.

What Is the Moon Phase and How Will It Affect Your Imaging Session?

For esthetic imaging projects, you likely want the darkest sky conditions with little or no moonlight. If you use a direct remote-control telescope, you can reserve a block of time when the Moon is not in the sky. During the days around a full Moon, you'll have little or no time at night without bright moonlight. However, the days closer to the first and last quarter Moon phase give you a window of time toward the beginning or end of a night without the Moon above the horizon. If you know the coordinates of the object or field of view you want to image and know when the Moon will set or rise, you can determine the best time to schedule your imaging run. Using a planetarium program to help plan your session times is a good way to ensure the times are right when you go to observe. Make sure you set the location in your planetarium program to the observatory location.

If you are using a queue-based remote telescope, you can submit your schedules for a day closer to a new Moon. Queue-based systems typically run your schedule the day you submit it. However, if your images were not taken owing to bad weather or a technical problem, by default they usually stay in the schedule queue to run on the next available night. If too much time passes without your schedules being able to run, you might have to delete them from the queue and reschedule for a better opportunity. Queue-based systems also enable you to create schedules that you set to run on a specific date only. If the schedule doesn't run on the specified night for any reason, you must resubmit a new schedule.

For astrometry programs in which you are trying to detect very faint objects near a telescope's limiting magnitude, such as asteroid discovery programs, you likely want to schedule your images for times when moonlight will not affect them.

Moonlight is not as big a detriment for many photometry programs. Although moonlight lowers the SNR, you can still get excellent image data for differential photometry if the object is 40 degrees or more from the Moon. If the Moon is close to full, you should probably limit your images to objects 90 degrees or more away from the Moon and select brighter objects that will give you a better SNR.

How Will the Weather Affect Your Ability to Get Your Image Data?

Unless the remote observatory you choose to use is located in the Atacama Desert in Chile, clouds and storm systems will be an intermittent problem you need to monitor. Most observatory sites have some periods of bad weather. For example, sites in Arizona and New Mexico have mostly clear skies for much of the year. However, from mid-June to early September, these areas get most of their precipitation from a tropical monsoon condition that sets up during that time of year. Huge thunderstorms develop that, in addition to clouds and rain, can produce violent lightning strikes. The threat of lightning is so great that some observatories close down for much of the monsoon season to protect the valuable and sensitive observatory instruments from damage (Fig. 7.4).

Fig. 7.4 Mount Lemmon SkyCenter domes 9157 ft (2791 m) above sea level (Credit Adam Block)

Commercial remote observatories typically have links on their websites to enable you to check out the weather forecast for each site and local monitoring devices to give you up-to-the-minute local conditions. You can check the weather forecast for the observatory you want to use before you schedule a direct-control session or submit a schedule to a queue-based system. If the commercial provider has telescopes in different regions around the world, you have a better chance of finding one that has a good probability of clear skies on the day you want to get your image data. With a direct-control telescope you might be able to make your go/no-go decision to run at the last minute. Queue-based systems normally run your image schedules on the day you submit them, weather permitting. If the weather turns out bad and your schedule cannot run, it typically stays in the queue to run the next night unless you cancel it.

Even when the weather is clear, the seeing conditions can vary considerably from night to night at any given location. A site whose typical average seeing conditions are 1.5 arc-seconds might suffer from seeing conditions of 5 arc-seconds or worse on some nights. The Clear Sky Chart website (cleardarksky.com/csk/) calculates the seeing and transparency forecasts for most observatory sites. It may not give you the exact true conditions, but it will give you a good relative idea whether the seeing is likely to be good or bad. Seeing degrades when the upper-level winds become turbulent. A cold front passing after a storm system might produce very transparent skies but also poor seeing. On the other hand, a night with a high pressure system overhead might have mediocre transparency, but excellent seeing conditions. In general, if the jet stream is nearby or overhead of a site, the seeing is probably not going to be as good as it could be.

Is Your Object of Interest Well Placed to Image?

If you are new to astronomical imaging and just starting out, you may not know when the object you want to image is well placed in the sky. You might want to image M42 because you've see so many beautiful images of it and you want to own one of your own. However, if it is July, you'll have to wait many months until Orion is high in the sky between December and February. Ideally, you want to image your target when it is highest above the horizon looking through the least amount of air mass. This is particularly true if you want to get the absolute best quality images for an esthetic imaging program. However, if you are doing a photometry program on a variable star, you might decide to image it even though it will be low in the sky just before or after twilight because the data are important for your program.

If an object with a southerly declination is located very low in the south or never visible from a Northern Hemisphere observatory, it might be located high in the sky at a Southern Hemisphere observatory. Choosing an observatory in the proper hemisphere is important to image objects near the northern or southern celestial poles.

Making the Most Efficient Use of the Remote Observatory

Once you have a solid, workable observing plan, you will want to implement it in the most efficient way possible. Doing so will save you time and money. If you decide to operate a direct remote-control telescope for your program, have your plan ready and use some method to keep notes of your session. Keeping notes will speed up your learning curve and maybe keep you from repeating a mistake or doing something inefficiently.

If you are charged for the time you set up and operate a remote telescope, you don't want to waste valuable time. Queue-based schedules enable you to take your time and double check everything before you submit your schedules. If things don't work out exactly as you hoped, make the necessary adjustments and revise your plan to make it more efficient.

Get to know the observatory owner/operator. He or she can be a great help in resolving issues and in learning any little quirks the observatory system may have. Learn about the local climate around the observatory and use any Internet resources to monitor the weather and how it may affect your observations. The owner/operator of the observatory will have a lot of information about the observatory systems, so take advantage of that resource.

Further Reading

Arditti D (2007) Setting-up a small observatory. Springer
Berry R, Burnell J (2005) The handbook of astronomical image processing. Willmann-Bell
Buchheim R (2007) The sky is your laboratory. Springer
Byrne CJ (2005) Lunar Orbiter photographic atlas of the near side of the Moon. Springer
Chromey FR (2010) To measure the sky. Cambridge
Covington MA (1999) Astrophotography for the amateur. Cambridge
Dragesco J (1995) High resolution astrophotography. Cambridge
Dymock R (2010) Asteroids and dwarf planets and how to observe them. Springer
Harrison KM (2011) Astronomical spectroscopy for amateurs. Springer
Henden AA, Kaitchuck, RH (1990) Astronomical photometry. Willmann-Bell
Howell SB (2006) Handbook of CCD astronomy. Cambridge
Hubbell GR (2012) Scientific astrophotography: how amateurs can generate and use professional
      imaging data. Springer
Shirao M, Wood CA (2010) The Kaguya lunar atlas. Springer
Smith GH, Ceragioli R, Berry R (2012) Telescopes, eyepieces and astrographs. Willmann-Bell
Warner BD (2006) A practical guide to lightcurve photometry and analysis. Springer
Warner BD (2010) The MPO user's guide. BDW Publishing

Websites

Cherry Mountain Observatory, www.cherrymountainobservatory.com
iTelescope, www.itelescope.net
LightBuckets, www.lightbuckets.com
New Mexico Skies, www.nmskies.com
Sierra Remote Observatories, www.sierra-remote.com
Sierra Stars Observatory Network (SSON), www.sierrastars.com
Slooh, main.slooh.com
University of Iowa Robotic Observatory (Rigel), astro.physics.uiowa.edu/rigel
University of Arizona Mt. Lemmon SkyCenter, Skycenter.arizona.edu
Warrumbungle Observatory, www.tenbyobservatory.com
Winer Observatory, www.winer.org
ADM Accessories, http://admaccessories.com/
Apogee Imaging Systems, http://www.ccd.com/
Astro Tech, https://www.astronomytechnologies.com/
Astro Hutech/BORG, http://www.sciencecenter.net/hutech/
Astro-Physics, Inc., http://www.astro-physics.com/
Atik Cameras, http://www.atik-usa.com/
Celestron, http://www.celestron.com/
Daystar Filters, http://www.daystarfilters.com/
Denkmeier Optical, Inc., http://www.deepskybinoviewer.com/
Explora-Dome (Poly Tank), http://www.exploradome.us/
Explore Scientific LLC, http://www.explorescientific.com/
Finger Lakes Instrumentation LLC, http://www.fli-cam.com/
Hotech Corp, http://www.hotechusa.com/
Howie Glatter's Laser Collimators, http://www.collimator.com/
Innovations Foresight LLC, http://www.innovationsforesight.com/
IOptron, http://www.ioptron.com/
Lunt Solar Systems LLC, http://www.luntsolarsystems.com/
Meade Instruments, http://www.meade.com/

Moonlite Telescope Accessories, http://www.focuser.com/
Peterson Engineering Corp, http://www.petersonengineering.com/sky/index.htm
Planewave Instruments, http://www.planewaveinstruments.com/
QHYCCD (Astro Factors/Deep Space Products), http://www.astrofactors.com
QSI, http://qsimaging.com/
Questar, http://www.questar-corp.com/
Santa Barbara Instrument Group, http://www.sbig.com/
Shelyak Instruments, http://pagesperso-orange.fr/shelyak/en/index.html
Sky Watcher, http://www.skywatcher.com/
Software Bisque, http://www.bisque.com/
Starlight Instruments LLC, http://www.starlightinstruments.com/
Starlight Xpress Ltd., http://www.sxccd.com/
Stellarvue, http://www.stellarvue.com/
Tele Vue Optics, http://www.televue.com/
Texas Nautical/Takahashi America, http://www.takahashiamerica.com/
Vernonscope, http://www.vernonscope.com/
Vixen Optics, http://www.vixenoptics.com/
William Optics, http://www.williamoptics.com/

Chapter 8

Picking Targets and Scheduling Your Observations

"So Much Data, So Little Time"

If you are fortunate enough to own and operate your own remote observatory, you have the entire night to use as you please. In contrast, professional institutional and commercial observatories often have a high demand for imaging time during the dark nights on either side of a new Moon. These facilities are valuable assets with a finite capacity that many people want to use for their astronomy programs. Some factors limit how much observing time is available for imaging, such as the total number of hours of darkness and inclement weather. During the summer months in the mid-latitudes of the Northern and Southern Hemispheres, there may be as little as 6 h of total dark time to use. In the winter, these same observatories may have 12 h or more of dark time. Poor weather conditions can keep an observatory closed for days or weeks. Having access to observatories on different continents in both hemispheres gives you a much better chance to get the image data you want on any particular night.

Because an observatory's time is so limited and valuable, it makes sense to optimize its use as much as possible. As described in previous chapters, queue-based observatory operations can handle taking many more images for a larger number of people than a direct-control remote system can. For sheer efficiency in getting the most data in a night with a remote observatory, a queue-based scheduling system is the best option.

© Springer International Publishing Switzerland 2015
G.R. Hubbell et al., *Remote Observatories for Amateur Astronomers*, The Patrick Moore Practical Astronomy Series, DOI 10.1007/978-3-319-21906-6_8

Remote Observatory Scheduling System Types

To schedule the use of a direct-control telescope, you typically check to see which one is available, chose it, and start operating it. The enterprise may also allow you to book (reserve) a telescope for use on a specified date and block of time. Operating a telescope immediately when you want to is the quickest way to get your image data. However, if the telescope you want to use is not available at the time you want to use it, you may have to wait for a time that is less opportune or you may miss the opportunity altogether if an astronomical event occurs at a specific time.

Queue-based scheduling systems allow you to set up the parameters of your image schedules and let the system handle the rest for you. They also give you more flexibility if your program requires precise time intervals between images in your schedule request that spans several hours.

Remote Observing in Batch Mode to Get the Most Data for the Least Cost

Queued scheduling is sometimes referred to as running in batch mode. Batch mode is a method that takes a collection of schedule requests and creates a master schedule to run during an entire night. A software program arranges the timing of the image exposures to run in an optimal sequence. Ideally, by default, the algorithms (rules) of these scheduling programs place each image (or series of images) in the schedule when the object of interest is highest in the sky near the Meridian for the best seeing conditions. However, if there are a large number of images in an observing run that take up most of the available time, conflicts for the same optimal times occur.

Solving this problem is much more difficult than it first appears and can challenge even the most sophisticated sorting algorithms. A large number of schedule requests may be received for objects in close proximity in the sky when they are all in an optimal position around the same time — these compete for the same time slot. The same observing run might also include schedule requests for a series of images with fixed time intervals that span several hours. These images have to interleave among the other requests in the schedule. To help resolve these conflicts, most scheduling software programs enable the observatory managers to set the priority of schedule requests, giving certain schedule requests precedence to run before schedule requests with a lower priority setting competing for the same time.

The most demanding schedule requests are the ones that need to start at a specific, fixed time and take images contiguously for a block of time. For example, if you want to determine the exact minimum of an eclipsing cataclysmic binary star system with a known period, you want your schedule request to start a certain amount of time before the projected time of the minimum and run for a certain amount of time afterward to ensure you catch the entire event. In addition, you likely want your images to run back to back throughout the process to get the best resolution of the resulting light curve and determine the exact time of the minimum with the highest precision possible. Obviously, for such a schedule to be successful, it must have a higher priority than any competing schedule in the observing run. This type of schedule request can have a major impact on the rest of the schedule because it does not allow the interleaving of images for other schedule requests between the beginning and end time of the schedule. Therefore, you may have to make a special arrangement with the observatory managers in advance for them to run these types of schedules.

Queue-based schedule systems enable many users to schedule images in an observing run in an efficient manner. Images are arranged in the most optimal position in the observing run without you having to do anything extra. More people can share the use of valuable observatory time in a given night. This compaction of many schedule requests into an efficient overall master schedule also enables commercial remote observatory companies to serve more users at a reasonable cost while realizing a profitable return on their investment.

The Information Needed to Schedule Observations

Before you command a direct-control remote telescope or set your schedule to run on a queue-based remote telescope, you should have a solid plan as discussed in Chap. 7. The following sections describe things you need to know to schedule successful observations.

What Do You Want to Image?

Commercial remote observatory facilities make it easy to select an object to image. You typically use a menu or form to select an object from catalog databases that may contain hundreds of thousands of objects. Sometimes you may want to image a specific field of view (FOV) instead of selecting an object. This is the case for programs such as asteroid discovery work in which you want to point to and track on a specific set of right ascension and declination coordinates to find an undiscovered asteroid or comet. The basic technique to discover new asteroids is to take a series of images over an hour or longer time period at coordinates you set and to use a software program to detect any objects that move between each of the images in the series.

If your project goal is to collect images to create a great esthetically pleasing image, you can select among thousands of objects from online catalogs of the remote observatory's online system. To get the best results, you should chose telescopes that match the magnitude (brightness) and size (in arc-minutes) of the object you plan to image. You can image a bright, large object such as the Orion Nebula (M42) with a small-diameter telescope (less than 16 in. (0.4 m)) in a reasonably short series of exposures to get excellent results. Because these size telescopes have a wide FOV, they enable you to capture the entire nebula and the surrounding area, which can produce stunning results. Alternately, you might choose to image the

Fig. 8.1 Image of M42 taken with a 5-in. (0.13-m) wide-field refractor at Mount Lemmon, Arizona (Credit Adam Block)

inner details of the nebula around the Trapezium part of the nebula. For this perspective, you would use a larger diameter, longer focal length telescope that enables you to achieve a larger image scale. Figures 8.1 and 8.2 show examples of images of the Orion Nebula (M42) created with image data from a 5-in. (0.13-m) wide-field refractor and a 24-in. (0.61-m) Cassegrain telescope taken by Adam Block from the summit of Mount Lemmon in Arizona. Both direct-control and queue-based remote observatory systems work well for doing these types of projects.

If your program goal is to get precise photometry measurements of an asteroid over an extended period of time to determine its rotation period, you should choose a telescope that can attain a desired signal-to-noise ratio (SNR) in a relatively short exposure time for the filters you use. This type of program requires you to collect many images during an observing run. Ideally, you want to get as many data points as possible during an observing period to create a plot of the varying magnitude measurements as the asteroid rotates or tumbles in its orbit. Because asteroid rotation periods can range from a couple of hours to more than 24 h, you likely need to get data from many nights to ensure you get the true

Fig. 8.2 Image of M42 taken with a 24-in. (0.61-m) Cassegrain telescope at Mount Lemmon, Arizona (Credit Adam Block)

period. When an asteroid is near opposition, it will be observable throughout an entire observing run. For this type of program, you want to get as many measurements as closely spaced as is practical for hours at a time and to repeat the process over a span of a few to several nights. Obviously, operating a direct-control remote telescope for this type of program would require your attention for hours at a time, which would be tedious at best. Also, operating this way might occupy a telescope for an entire observing run.

Queue-based remote observatories are ideal for long repetitive observing program that require getting lots of data over long periods to collect enough data for light curve measurements to determine a precise rotation period of an asteroid. You can create schedules to image an asteroid every few minutes for several hours over many nights without the need for you to "be there" in the process (Fig. 8.3).

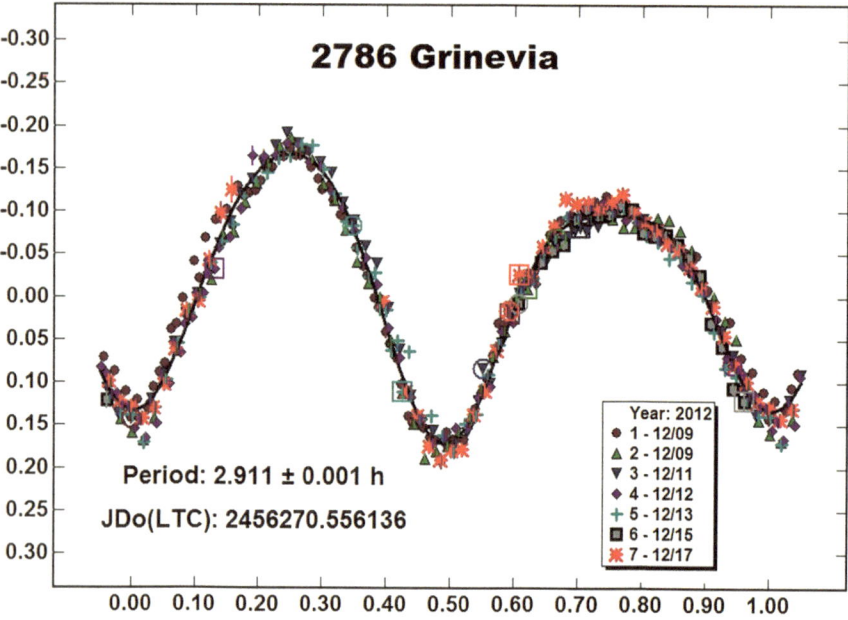

Fig. 8.3 Asteroid light curve plot of asteroid 2786 Grinevia from data taken by New Mexico Tech, Magdalena Observatory (Credit Daniel A. Klinglesmith III, Jesse Hanowell, Ethan Risley, Janek Turk, Angelica Vargas, Curtis Alan Warren Etscorn Campus Observatory, New Mexico Tech)

What Is the Optimum Time in an Observing Run to Take Your Images?

Unless the object you want to image is circumpolar and never sets at the observatory site you use, it will be below the horizon and not visible during some months of the year and hours of the night. The optimum time to take images of an object is when it is highest in the sky (crosses the Meridian) with the least amount of air to look through. The latitude of an observatory site determines how high in the sky an object will be when it crosses the Meridian. For example, the star Vega in the constellation Lyra passes directly overhead near the zenith when it crosses the Meridian in mid-northern latitudes, but it will just barely rise above the horizon in mid-southern latitudes. Obviously, you would choose an observatory in the Northern Hemisphere to image objects in the vicinity of Vega such as M1, the Crab Nebula.

As the Earth moves around the Sun throughout the year, different constellations reach their highest point in the sky during an observing run. Six months later, these same constellations disappear from view in the glare of the Sun. You must determine whether the objects you want to image will be visible and, ideally, are high in the sky during an observing run. You can check this quickly using monthly or seasonal star charts.

In mid-latitudes during the summer months, the total number of observing hours between morning and evening astronomical twilight may be only about 6 h. In the winter at these same sites, the total number of observing hours may be more than 12 h.

If you use a direct-control remote telescope, you want to know beforehand when the best time is to gain access to the telescope. The best time could be offset by as much as 12 h from the local time where you plan to operate the telescope.

Queue-based remote telescope systems try to put your observing requests in the queue as close as possible to when your object is highest in the sky by default. Even so, you must be aware of where your object of interest is located and whether it will rise high enough during an observing run to meet your needs. If your object sets early or rises late in an observing run, a queue-based system will try to place your schedule requests at the highest position for the observing run. However, it may be tens of degrees lower in the sky than it would be if you scheduled the observation at a different time of the year.

How Long Should Your Exposures Be?

The best exposure times to set for your images depend to a large degree on what it is you want to accomplish with your observing program. If your goal is to create a beautiful high-definition, high-contrast color composite image of a faint nebula or galaxy, then your program may require multiple long exposures using several different filters to get the image data you need to achieve the result you seek. For this type of program, you probably want to acquire as much total exposure time as your budget allows. However, the longest actual exposure times of your individual images will depend on the capabilities and limitations of the systems you use.

A direct-control telescope may have an auto-guiding system you can set up to track on a guide star for long periods of time. For these systems, you might be able to track well enough to get exposure times of 30 min or more without interruption. Such long exposures can produce results with high contrast and a high SNR. The risk of these long exposure times is that something unforeseen, such as a meteor or aircraft light trails or a jump in the tracking for which the auto-guider cannot compensate, might spoil the exposure, and the time and effort will be wasted. In addition, if your image is in a field with very bright stars near the faint detail you want to bring out, the bright stars might overexpose, flooding light into surrounding pixels causing a bloom effect, which might detract from the result you want to achieve. The solution in these cases is to take images with shorter exposure times and stack (add) them together to create an equivalent total exposure time.

Queue-based remote systems typically run unguided to make the most efficient use of the observing time available. These systems have high-precision tracking capabilities that enable them to track on an object for 5–10 min or more without the need to auto-guide. Therefore, these systems typically limit the maximum exposure times you can set for your images to about 300 s (5 min) to ensure high-precision tracking. The high-performance charge-coupled device (CCD) cameras these systems use produce a small amount of electronic and thermal noise as they cool

Fig. 8.4 A single 180-s exposure time unfiltered image of the Crab Nebula (M1) taken with the Mount Lemmon SkyCenter 32-in. (0.81-m) telescope (Credit Adam Block)

the CCD chip to 50–65 °C below the ambient temperature. Adding images to create longer effective exposure times increases the SNR without adding much additional noise from electronic or thermal effects. For photometry and astrometry projects, your main concern is what SNR you want to achieve for your project goals.

Figure 8.4 shows a calibrated unfiltered 180-s exposure of the Crab Nebula (M1) taken with the Mount Lemmon SkyCenter 32-in. (0.8-m) telescope. Even with such a short exposure, you can see a wealth of detail. Figure 8.5 shows the result you can achieve by combining ten 180-s exposures of M1 taken with the same telescope on the same night to create an image with a total effective exposure time of 30 min. The SNR, contrast, and detail are enhanced. The images are cropped around the nebula to increase the image scale to show more detail.

What Is Your Budget?

Most commercial remote astronomy companies charge you upfront to use their services. They use two different pricing methods. For example, iTelescope uses a membership plan in which you pay a recurring monthly fee ranging from $20 AUD to $1000 AUD. Unused money accumulates in your account. Sierra Stars Observatory Network (SSON) and LightBuckets enable you to set up an account

Fig. 8.5 A composite (stacked) image of ten 180-s exposure time unfiltered images of the Crab Nebula (M1) taken with the Mount Lemmon SkyCenter 32-in. (0.81-m) telescope (Credit Adam Block)

and pay any amount you want for credits (SSON) or points (LightBuckets) with a cost of $1 USD per credit or point. There is no monthly recurring fee for these systems. Each company offers discounts for larger purchases.

All these companies charge different rates in credits/points per hour for tele-scope time, with larger telescopes normally costing more than smaller ones. Starting in 2007, SSON was the first company to charge for only the actual exposure times used without any additional overhead costs SSON was also the first to offer a 100 % guarantee that you are completely satisfied with the quality of the image data you receive or you are reimbursed for the credits you used. Now, all these companies follow this model, which is a great service to everyone using a commercial remote observatory company.

Computer planetarium software and online planetarium programs are great tools to help you plan your remote astronomy observations. A simple view of the current constellations visible from a specified location, date, and time give you valuable information about whether an object you want to image is well placed for a remote observatory site. The more advanced planetarium programs have databases with up to millions of stars, galaxies, and nebula you can plot and zoom in on. These planetarium programs also allow you to download and install the latest, up-to-date minor planet catalog from the Minor Planet Center to determine the exact location of a given asteroid you may be interested in at a specific time. Whether your observing goals are simple or complex, a planetarium program enables you to analyze many "what if" opportunities for observing the current sky or any date well into the future.

You can also use a planetarium program to determine the phase of the Moon and how far it will be from an object you are interested in observing on a given night. This can help you make an informed decision about whether or when to observe an object that might be affected by moonlight during part or all of an observing run.

Dealing with the Weather

Every observatory site experiences periods of poor weather conditions. When planning your imaging sessions, you should check the local weather forecast for the observatory site you want to use. The National Oceanic and Atmospheric Administration (NOAA) (United States) and other national weather agencies give you detailed information about the current weather conditions for virtually any city and town around the world and long-range weather forecasts for the next 5–10 days. Their websites tell you the expected amount of cloudiness and projected temperatures, humidity, and chance of precipitation. In addition, most of them provide satellite images of the current and recent past cloud cover in the visual and infrared bandwidth. You can use this information to see whether the observatory site you want to use will have a reasonable chance of good observing conditions on the night you want to schedule your imaging run. If you find that the weather will be poor at one observatory site, you may be able to use a different observatory site for that night. Many professional observatory facilities also provide real-time monitoring of the local weather conditions with weather monitoring systems and web cameras.

Further Reading

Arditti D (2007) Setting-up a small observatory. Springer
Berry R, Burnell J (2005) The handbook of astronomical image processing. Willmann-Bell
Buchheim R (2007) The sky is your laboratory. Springer
Byrne CJ (2005) Lunar Orbiter photographic atlas of the near side of the Moon. Springer
Chromey FR (2010) To measure the sky. Cambridge
Covington MA (1999) Astrophotography for the amateur. Cambridge
Dragesco J (1995) High resolution astrophotography. Cambridge
Dymock R (2010) Asteroids and dwarf planets and how to observe them. Springer
Harrison KM (2011) Astronomical spectroscopy for amateurs. Springer
Henden AA, Kaitchuck, RH (1990) Astronomical photometry. Willmann-Bell
Howell SB (2006) Handbook of CCD astronomy. Cambridge
Hubbell GR (2012) Scientific astrophotography: how amateurs can generate and use professional
    imaging data. Springer
Shirao M, Wood CA (2010) The Kaguya lunar atlas. Springer

Smith GH, Ceragioli R, Berry R (2012) Telescopes, eyepieces and astrographs. Willmann-Bell
Warner BD (2006) A practical guide to lightcurve photometry and analysis. Springer
Warner BD (2010) The MPO user's guide. BDW Publishing

Websites

Cherry Mountain Observatory, www.cherrymountainobservatory.com
iTelescope, www.itelescope.net
LightBuckets, www.lightbuckets.com
New Mexico Skies, www.nmskies.com
Sierra Remote Observatories, www.sierra-remote.com
Sierra Stars Observatory Network (SSON), www.sierrastars.com
Slooh, main.slooh.com
University of Iowa Robotic Observatory (Rigel), astro.physics.uiowa.edu/rigel
University of Arizona Mt. Lemmon SkyCenter, Skycenter.arizona.edu
Warrumbungle Observatory, www.tenbyobservatory.com
Winer Observatory, www.winer.org
ADM Accessories, http://admaccessories.com/
Apogee Imaging Systems, http://www.ccd.com/
Astro Tech, https://www.astronomytechnologies.com/
Astro Hutech/BORG, http://www.sciencecenter.net/hutech/
Astro-Physics, Inc., http://www.astro-physics.com/
Atik Cameras, http://www.atik-usa.com/
Celestron, http://www.celestron.com/
Daystar Filters, http://www.daystarfilters.com/
Denkmeier Optical, Inc., http://www.deepskybinoviewer.com/
Explora-Dome (Poly Tank), http://www.exploradome.us/
Explore Scientific LLC, http://www.explorescientific.com/
Finger Lakes Instrumentation LLC, http://www.fli-cam.com/
Hotech Corp, http://www.hotechusa.com/
Howie Glatter's Laser Collimators, http://www.collimator.com/
Innovations Foresight LLC, http://www.innovationsforesight.com/
IOptron, http://www.ioptron.com/
Lunt Solar Systems LLC, http://www.luntsolarsystems.com/
Meade Instruments, http://www.meade.com/
Moonlite Telescope Accessories, http://www.focuser.com/
Peterson Engineering Corp, http://www.petersonengineering.com/sky/index.htm
Planewave Instruments, http://www.planewaveinstruments.com/
QHYCCD (Astro Factors/Deep Space Products), http://www.astrofactors.com
QSI, http://qsimaging.com/
Questar, http://www.questar-corp.com/
Santa Barbara Instrument Group, http://www.sbig.com/
Shelyak Instruments, http://pagesperso-orange.fr/shelyak/en/index.html
Sky Watcher, http://www.skywatcher.com/
Software Bisque, http://www.bisque.com/
Starlight Instruments LLC, http://www.starlightinstruments.com/
Starlight Xpress Ltd., http://www.sxccd.com/
Stellarvue, http://www.stellarvue.com/
Tele Vue Optics, http://www.televue.com/
Texas Nautical/Takahashi America, http://www.takahashiamerica.com/
Vernonscope, http://www.vernonscope.com/
Vixen Optics, http://www.vixenoptics.com/
William Optics, http://www.williamoptics.com/

Chapter 9

Remote Observing Projects for the Amateur Astronomer

Astronomy is a broad, fascinating subject to study. Your interests may compel you to do a project that contributes to the scientific advancement of astronomy in some way or you may be more interested in capturing the intricate beauty in the colors and structure of nebula, galaxies, and star fields. Whatever your interest and goal, a remote astronomy facility can make getting your image data easier and more pleasurable. The following sections describe different types of astronomy projects amateurs and professionals are doing today using remote observatory facilities.

Astronomical photometry is a technique to measure the brightness and color of stars, galaxies, nebula, and solar systems objects such as asteroids and comets. Except for samples of material collected by robotic spacecraft (and people walking on the Moon) and meteorites that land on Earth, everything we know about the universe outside of Earth comes from analyzing light emitted from or reflected off objects. With the exception of a few giant, relatively nearby (by astronomical standards) stars, even the largest telescopes see stars only as unresolved point sources of light. However, there is a wealth of information in the light of stars that you can capture with telescopes.

© Springer International Publishing Switzerland 2015
G.R. Hubbell et al., *Remote Observatories for Amateur Astronomers*, The Patrick Moore Practical Astronomy Series, DOI 10.1007/978-3-319-21906-6_9

The primary goal of photometry projects is to measure the brightness of objects as precisely as possible. In astronomy, you measure an object's brightness using a standard "magnitude" scale in which the magnitude number value increases as the brightness *decreases*. Each whole-number increase or decrease in magnitude represents a difference in brightness of approximately 2.512 times, resulting in a logarithmic scale. High-precision photometry measurements are recorded in fractions of a magnitude to a precision of three or more decimal places. The signal to noise ratio (SNR) of the object you measure in your image data determines how precise your magnitude measurements can be. The brighter the object you measure compared with the overall noise generated by electronic, thermal, and optical artifacts in the image data, the higher the SNR and, therefore, the precision of your photometric measurements. Even a small-diameter telescope can give you high SNR image data of bright stars in relatively short exposure times. Very dim objects may require you use a large-diameter telescope and long exposure times to achieve a desired SNR for your photometry measurements.

Astronomers use special photometry filters to determine the color characteristics of stars and other objects. The spectrum colors visible to the naked eye range from blue to red as you would see in a rainbow, with blue having the shortest wavelength and red having the longest wavelength. This is a very small part of the entire electromagnetic spectrum. Some charge-coupled device (CCD) chips can detect photons well into the ultraviolet and near infrared parts of the spectrum. The color of a star can tell you a lot about its temperature and its stellar classification. Color filters used for photometry block out all light except for a small part of the spectrum within the desired color range to be measured. The most common filters used for CCD photometry measurements are the Johnson-Cousins or Bessel-Johnson UBVRI set of filters, which stand for ultraviolet, blue, visual (green), red, and infrared, respectively (Fig. 9.1). The Johnson-Cousins photometry standard describes the width and wavelength that each filter will allow to pass while blocking light that is outside that bandwidth. Measurements taken by different people using different telescopes with standardized photometry filters can then be compared equally.

Variable Star Observing Projects

The American Association of Variable Star Observers (AAVSO) is the largest worldwide organization of people taking photometry magnitude measurements of variable stars. Several other organizations around the world also have an active variable star observing program. Many types of variable stars fluctuate in brightness for several different reasons. Some stars change brightness over long periods (months to years) while others change brightness in much shorter time frames (days or hours). Eruptive and cataclysmic variable stars can unpredictably change brightness by many magnitudes. The rarest types of variable stars are novae and

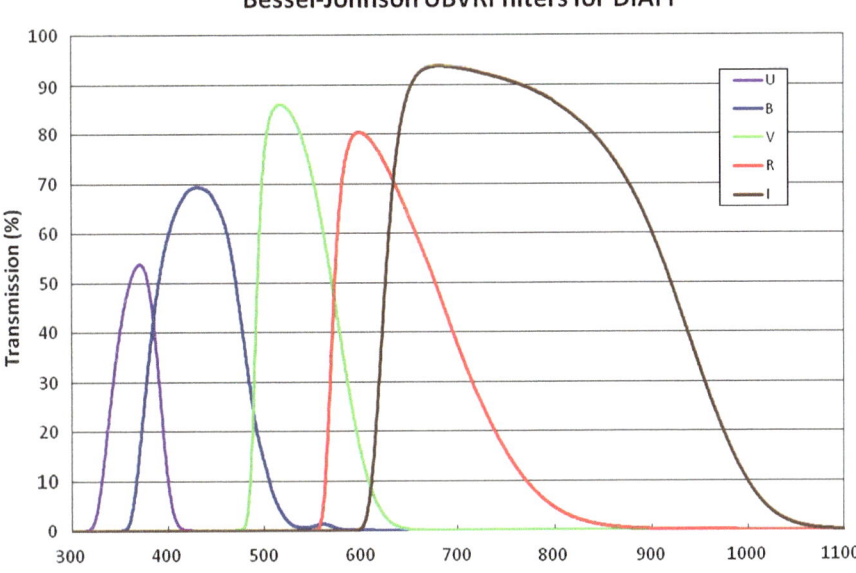

Fig. 9.1 UBVRI photometry filter passbands

supernovae. Their appearance is unpredictable but they can brighten quickly, increasing in brightness by 7 to more than 20 magnitudes in a short period of time.

One goal of variable star monitoring programs is to collect enough high-precision photometry measurements over time to determine an accurate light curve plot of the period. The magnitude measurement points in a plot show the variation in brightness over time. The more measurement points you have in a light curve plot throughout a period, the more precise and smooth the resulting plot curve will be.

In most cases, you can monitor long-period variable stars by measuring brightness once every day or so to generate an accurate light curve (Fig. 9.2). As more people contribute measurements over time, the precision and smoothness of the resulting light curve plot increases. Long-period variable stars have periods that range from about a month to a few years.

Short-period variable stars require more frequent monitoring to ensure that you properly resolve the correct light curve (Fig. 9.3). These types of stars can have periods ranging from as short as an hour to a few days or more. In some cases, such as for RR Lyrae stars, you can get photometry measurements that cover a few complete periods in a single evening. Of course, to ensure that you get a precise, smooth light curve plot of such short periods, you must acquire many images more or less back to back in a short amount of time.

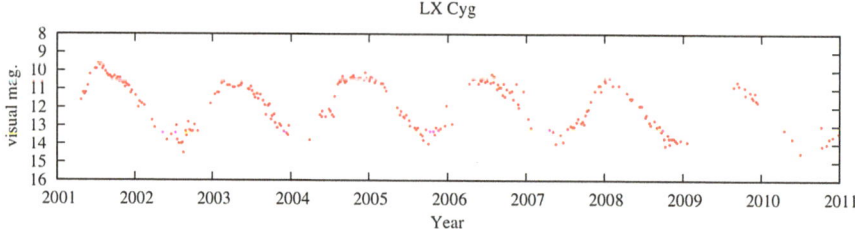

Fig. 9.2 Light curve graph of a long-period variable star from AAVSO data

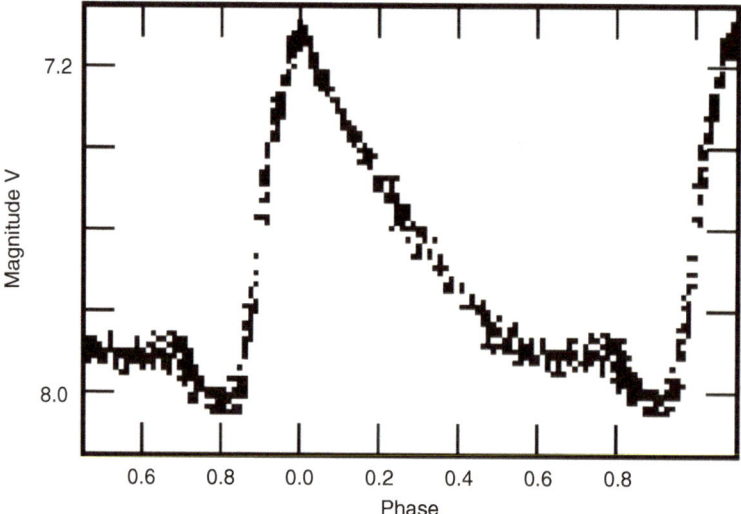

Fig. 9.3 Light curve graph of a short-period Delta Scuti variable star from NSVS archival data

Cataclysmic variable stars are binary star systems that have highly irregular periods. The binary system consists of a white dwarf star and another star with an orbit very close to the white dwarf. The companion star sheds some of its matter to the white dwarf because of the strong gravitational attraction between them. The matter accretes (collects) on the surface of the white dwarf until it reaches a critical mass, causing a runaway fusion reaction that results in a powerful thermonuclear explosion. These dramatic outbursts are unpredictable and can occur at any time. Consequently, cataclysmic variable stars require constant, regular monitoring to catch them in the act of a violent outburst. A few international groups of professional and amateur scientists specialize in monitoring these stars and have a quick-action communications network on the Internet to alert members to events and to coordinate the gathering of photometry measurements.

Remote observatory facilities are ideal tools for doing variable star observation projects. You can use direct-control remote telescopes to observe a cataclysmic or transient event on the fly soon after the event occurs. Queue-based automated systems are best to use for regular observations of most variable stars.

A comprehensive description of doing variable star observations is beyond the scope of this book. You can refer to the references and website links at the end of this chapter for more information.

Asteroid Light Curve Projects

Most asteroids are irregularly shaped rocky objects ranging in diameter from hundreds of feet to dozens of miles. Their rate of rotation (spin) can vary from a couple of hours to more than a day. An asteroid's rotation can be a simple spin around a single axis or may be a more complex rotation resembling a tumble. You can determine the rotation period of asteroids by taking many images of an asteroid during an observing run, repeating this process over several nights, and plotting the changes in magnitude (brightness) over time (see Fig. 9.4).

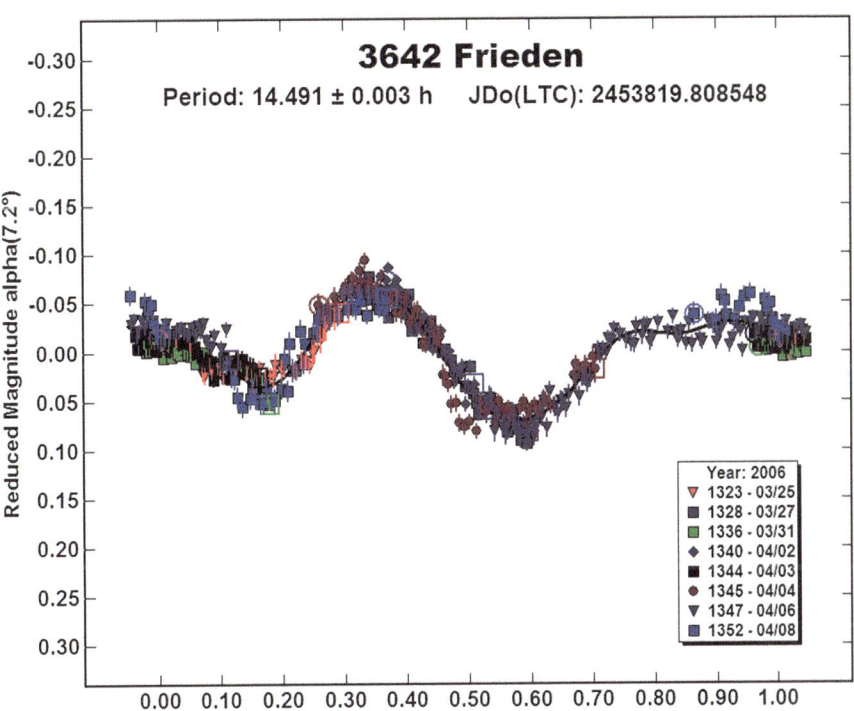

Fig. 9.4 Light curve graph of the Main Belt asteroid 3642 Frieden from data taken by Brian D. Warner at Palmer Divide Observatory

The general procedure to do an asteroid light curve project to determine an asteroid's rotation period is as follows:

1. Select an asteroid that will be bright enough to achieve photometry measurements with an SNR of 100 or better (with the filter you choose to use) to get high-precision results. Also select an asteroid that is near opposition to ensure that it will be well placed in the sky to image it for several hours during an observing run.
2. Set a schedule (or reserve a time slot) on the remote telescope you will use to take an image every 2 or 3 min for as many hours as is practical (or within your budget) for an observing run.
3. Use a software program such as MPO Canopus© or Peranso® that is designed to analyze asteroid image data to create light curve graphs and do complex period analysis.
4. Repeat steps 1–3 as many times as needed until you are satisfied that you have a good solution for the asteroid's rotation period.
5. Publish your period analysis and light curve results on the Minor Planet Bulletin operated by the Minor Planet Center (MPC).

Some asteroids have a gravitationally bound companion creating a binary system. You can determine whether this is the case by carefully analyzing your image data. Also, if you get enough image data over a period of months when an asteroid, Earth, and Sun are at different angles to each other (phase angle), you can get a good estimate of the asteroid's actual shape. You can use special software to create a 3D graphic model of an asteroid shape after you collect an appropriate amount of data over an extended period of time (Fig. 9.5).

Asteroid and Comet Astrometry Projects

Astrometry is a technique to measure the position of objects in the sky. Positions are measured using a Celestial Equatorial Coordinate System in which right ascension coordinates determine east/west positions and declination coordinates determine north/south positions. These form a grid similar to longitude and latitude coordinates used to designate terrestrial positions.

The positions of stationary objects, such as stars and galaxies, do not change significantly over the course of a human lifetime (although everything in the universe is moving relative to everything else). Therefore, their relative coordinate positions are essentially static. However, the Earth's rotational axis changes continually over time because the Earth wobbles like a spinning top from the gravitational pull of the Sun and Moon. Astronomers standardize the coordinates everyone uses by specifying an epoch date for the coordinates used. Standardized epoch dates are spaced 50 years apart, with the current latest epoch date as the year 2000 (usually designated as J2000). J2000 specifies a position in the coordinate system based on the Earth's rotational axis on January 1, 2000.

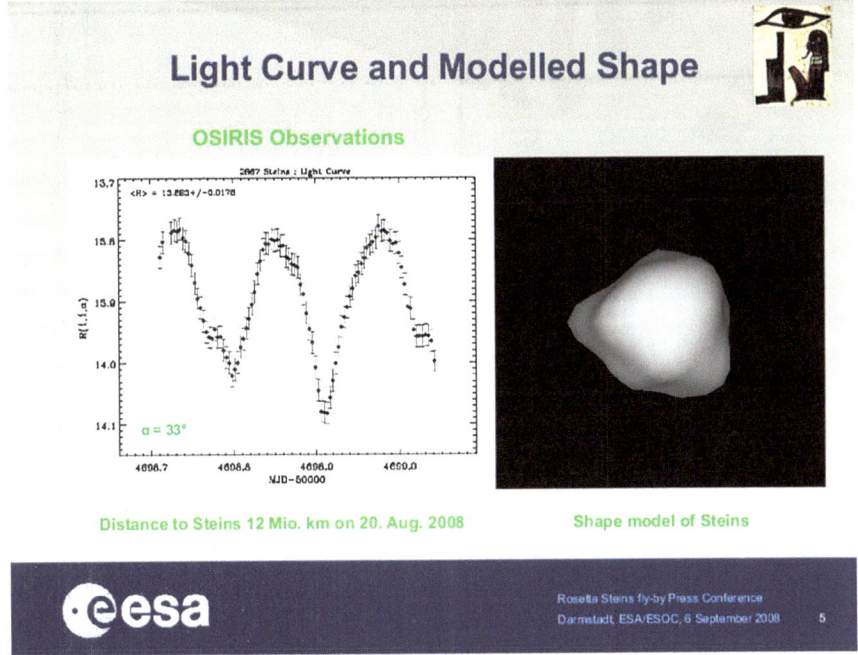

Fig. 9.5 Surface model of an asteroid created from observation data

Solar system objects such as asteroids and comets move noticeably against the relatively static star fields as seen from Earth. Their orbits around the Sun determine their position and path in the sky. In general, the closer a moving object is to Earth, the faster it appears to move against the sky background. Therefore, these objects move over time in the field of view of a telescope tracking at the sidereal tracking rate used to lock on to stationary objects such as stars and nebula. In long exposures, an asteroid leaves a trail in an image taken using a telescope tracking at the sidereal rate. To track on an asteroid or comet to keep it from trailing or smearing on the image, a telescope must track at the rate the object moves. This type of tracking requires the ability to set the telescope control system's tracking rate to the ephemeris and orbital elements of the target (Fig. 9.6).

The MPC is the central clearinghouse that creates ephemeris and orbital element data for asteroids and comets. It generates this information from precise astrometry position measurements sent to it by professional and amateur astronomers around the world. The MPC MPCORB database contains the orbital elements of hundreds of thousands of asteroids, with more added every day.

Hundreds of professional and amateur astronomers scan the sky on every clear dark night to search for near-Earth objects (NEO), primarily asteroids that might be a threat to impact the Earth and cause great damage. In the process of looking for NEOs, they mostly discover new Main Belt asteroids that orbit between Mars and

Fig. 9.6 Color composite image of comet 17P-Holmes from data taken with the Sierra Stars Observatory

Jupiter. You can discover new asteroids using remote telescope facilities. The general process is as follows:

1. Select a field of view in a set of coordinates within 10 degrees or so of the ecliptic. (Most asteroids orbit near the ecliptic, and the density degrees rapidly as you move north and south of it.)
2. Schedule a series three to four or more images with 120-s to 180-s exposure times to be taken at 10- to 20-min intervals with approximately an hour of time between the first and last exposure.
3. Download your completed images and use an astronomy image processing software program that can stack your images to create a movie loop that causes a moving object to "blink" as it moves from one frame (image) to the next in the sequence.
4. Use an image processing program (such as Astrometrica or MaxIm DL®) that can measure the precise right ascension and declination coordinates of any moving objects you find in the images and use the MPC's online MPChecker program to see what known asteroids are in the area of the moving object you found.

5. If there are no known asteroids in the vicinity of the moving object you found, it might be an undiscovered asteroid. Follow the MPC guidelines to send in the coordinates and precise time of the image exposures in the proper format to the MPC.
6. On the next possible night (preferably the night after your first observations) determine where the asteroid will be and follow steps 1–5 to send your observations to the MPC. You can get a good idea where an asteroid will be in the near future by figuring out the direction (the vector in the series of images) your object was moving and how fast it was moving in arc-seconds per hour.
7. If the MPC determines that your object is new and you are the first to find it or it is a new object that someone else found already, the MPC will send you an email notification.
8. After there are enough observations to determine a new asteroid's orbit to a high precision, MPC assigns it a number and the discoverer (You!) can name it. However, this process can take years to complete.

The above process is general. There are many finer points you need to be aware of to do this correctly and be successful. The MPC website has a thorough explanation and guide for doing asteroid discovery and follow-up observations.

Esthetic Imaging Projects

Many of the projects described in this book are about doing meaningful scientific observations. However, your goal may be to capture the beauty of the many types of celestial objects. You've most likely seen the fantastic color images taken with the Hubble Space Telescope and major Earth-based observatories, and by accomplished amateur astronomers using their own equipment. You can achieve similar beautiful results from images you acquire using remote observatories. To get the best results, you must start with high-quality images. No amount of image processing after the fact can make bad data good. So what makes a high-quality image? It's a combination of the quality of the instruments used and the surrounding environment at the time the images are taken (Fig. 9.7).

Instrument Concerns for High-Quality Images

Consider the following criteria when choosing a remote observatory for taking high-quality images:

- The remote telescope you use should have a high quantum efficiency camera with a pixel resolution matched to the image scale of the telescope.
- Raw images must be fully calibrated with appropriate flat field frames for each filter used to ensure that no optical imperfections project into the image data.
- The telescope must point and track with high precision throughout exposures to ensure that the intended object (or field of view) is well centered and that there is no elongation or trailing from tracking errors.

Fig. 9.7 Color composite image of NGC 2359 by Adam Block using the LRGB filter set on the Mount Lemmon SkyCenter 32-in. (0.81-m) telescope (Total exposure time: 20 h (L = 10, R = 4, G = 3, B = 3))

Environmental Concerns for High-Quality Images

Consider the following environmental criteria when choosing a remote observatory for taking high-quality images:

- Light pollution brightens the sky background and washes out the contrast in images. The remote observatory you select should be at a site with little or no manmade light pollution.
- Moonlight is another form of light pollution. Ideally, you want to schedule your imaging when the Moon is near its new phase and/or when the Moon is below the horizon.
- The closer your target is to the horizon, the more air mass your instruments have to see through, which can degrade the resolution in your images. Ideally, you want to collect your image data when your target is high in the sky. Therefore, you need to select objects that will be near transit (i.e., highest point in the sky

from a given location) at some point during the observing run and, if you have the option, choose a telescope located at a latitude that places the object highest in the sky.

- Seeing determines the resolution (sharpness of detail) in your raw image data. Some professional observatory sites are renowned for having consistently excellent local seeing conditions, with many nights having better than 1 arc-second seeing. Some remote observatory sites have better seeing than others on average. However, even the best locations can have extremely poor seeing at times, depending on wind speed and direction, location of the Jetstream overhead, weather fronts, and so on.
- High thin clouds (such a cirrus clouds) can attenuate and dissipate some of the light coming from your target. Although this might lower the SNR while still giving acceptable results for a photometry or astrometry project, it might spoil image data you want to use for your esthetic imaging project.

Filters for Esthetic Images

You create color images by combining greyscale CCD images taken using various filters. One of the most popular filter techniques used to create color composite images is called LRGB, which stands for Luminance, Red, Green, and Blue. The Luminance filter is a clear filter that blocks light in the near infrared, a part of the spectrum to which many CCD chips are sensitive. The purpose is to collect light across the spectrum to get as much intensity, contrast, and detail as you can of the target object. Using no filter (open slot) serves the same purpose. The Luminance image provides the "white light" base of a color composite image. You could take images with R, G, and B only, which pass light through a band corresponding to the red, green, and blue parts of the spectrum, respectively. However, although the RGB images create a blended, full-color image when combined, taking these images may require very long total exposure times to get high-contrast results and true white intensity. Using Luminance images in combination with RGB images can give you more pleasing results in shorter total exposure times (Fig. 9.8).

Other Project Ideas

Besides the project ideas already described in this chapter, there are others you might decide to try. The following list is by no means exhaustive, and you can probably come up with several more on your own.

- Monitoring and imaging variable nebulae
- Searching for and discovering new variable stars
- Searching for comets close to the Sun near the start and end of astronomical twilight
- Conducting spectroscopy studies of stars, nebulae, and planets (Fig. 9.9)

Fig. 9.8 Color composite image of M104 using the LRGB filter set on the Mount Lemmon SkyCenter 24-in. (0.61-m) telescope (Total exposure time: 310 min (L=210, R=30, G=30, B=40))

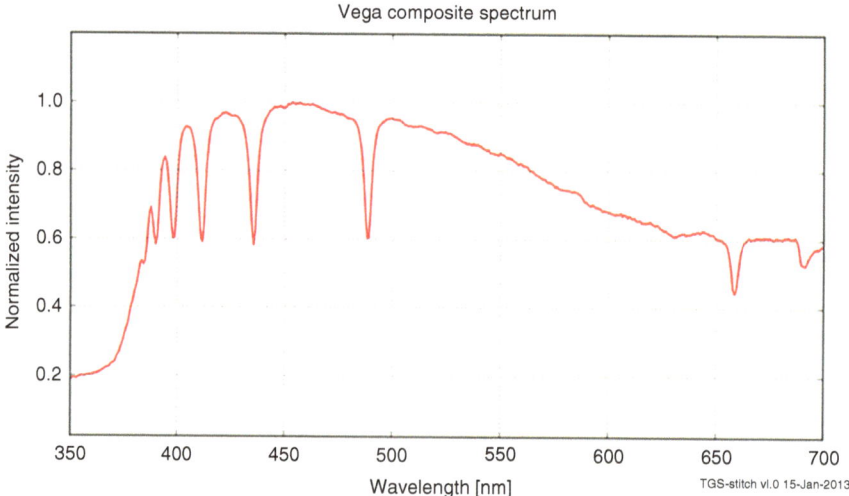

Fig. 9.9 Low-resolution spectrum plot of Vega created using data acquired remotely with the transmission grating spectrograph (TGS) on the University of Iowa Rigel Telescope in Sonorita, Arizona

Working on Projects with Groups and Organizations

While you can work by yourself on projects indefinitely, producing excellent data and even publishing your results in peer-reviewed journals, you might consider working on projects with other people. Working with a group has many advantages. Some people in a group might have decades of experience working on the same type of project that interests you. Group members are often passionate about their work and eager to help others get started in their field of expertise. If you participate in a group project and share your data, you might get the chance to have your name cited in a project's peer-reviewed paper in a professional journal.

If you do a Google® search online of a topic you are interested in investigating as a potential astronomy project, you'll likely find links to groups of people doing research projects on the subject. Some of the groups are made up of dedicated amateur astronomers (also referred to as citizen scientists) who work closely with professional astronomers in a mutual collaboration. Below is a list of some groups and organizations doing scientific studies on several different astronomy subjects that can help you get started finding a project you might want to get involved in. Many seek active participants and collaborators.

- *American Association of Variable Star Observers (AAVSO)*

    The AAVSO has thousands of members who are professional and amateur astronomers from around the world and who contribute nearly 1 million observations a year. The AAVSO website has a wealth of information—from how to get started to current specific sophisticated observing campaigns looking for people to collaborate. These campaigns have their own network of remote observatories that members can use to make their own observations.
- *British Astronomical Association (BAA)*

    The BAA has many observing section groups that study virtually every aspect of observational astronomy from the solar system to deep sky objects. It has a Robotic Telescope Project that provides members a 50 % subsidy to use commercial remote observatories for approved projects.
- *Center for Backyard Astrophysics (CBA)*

    The CBA is an international organization of observers with relatively small telescopes (8–26 in. in diameter) dedicated to the photometry study of cataclysmic variable stars. Members communicate in near-real time via email for coordinating activities on ongoing projects. Some members use commercial remote observatories for their observations.
- *International Astronomical Search Collaboration (IASC)*

    The IASC is an educational outreach program for high schools and colleges. It provides high-quality astronomical data to students around the world. Students can make original astronomical discoveries and participate in hands-on astronomy. The program uses remote observatories around the world.
- *Target Asteroids!*

    Target Asteroids! and Target NEOs! are citizen science projects of the NASA OSIRIS-REx asteroid sample return mission that provide data for continued study of near-Earth asteroids. Target Asteroids! is a communication and public engagement project of the OSIRIS-REx mission. Target NEOs! is a parallel observing program for Astronomical League members. Since 2012, observations from both projects have benefited the OSIRIS-REx Science Team and other researchers. The OSIRIS-REx Science Team uses the Sierra Stars Observatory Network (SSON) on a regular basis for asteroid observations.

Further Reading

Arditti D (2007) Setting-up a small observatory. Springer
Berry R, Burnell J (2005) The handbook of astronomical image processing. Willmann-Bell
Buchheim R (2007) The sky is your laboratory. Springer
Byrne CJ (2005) Lunar Orbiter photographic atlas of the near side of the Moon. Springer
Chromey FR (2010) To measure the sky. Cambridge
Covington MA (1999) Astrophotography for the amateur. Cambridge
Dragesco J (1995) High resolution astrophotography. Cambridge
Dymock R (2010) Asteroids and dwarf planets and how to observe them. Springer
Harrison KM (2011) Astronomical spectroscopy for amateurs. Springer
Henden AA, Kaitchuck, RH (1990) Astronomical photometry. Willmann-Bell
Howell SB (2006) Handbook of CCD astronomy. Cambridge
Hubbell GR (2012) Scientific astrophotography: how amateurs can generate and use professional imaging data. Springer
Shirao M, Wood CA (2010) The Kaguya lunar atlas. Springer
Smith GH, Ceragioli R, Berry R (2012) Telescopes, eyepieces and astrographs. Willmann-Bell
Warner BD (2006) A practical guide to lightcurve photometry and analysis. Springer
Warner BD (2010) The MPO user's guide. BDW Publishing

Websites

Cherry Mountain Observatory, www.cherrymountainobservatory.com
iTelescope, www.itelescope.net
LightBuckets, www.lightbuckets.com
New Mexico Skies, www.nmskies.com
Sierra Remote Observatories, www.sierra-remote.com
Sierra Stars Observatory Network (SSON), www.sierrastars.com
Slooh, main.slooh.com
University of Iowa Robotic Observatory (Rigel), astro.physics.uiowa.edu/rigel
University of Arizona Mt. Lemmon SkyCenter, Skycenter.arizona.edu
Warrumbungle Observatory, www.tenbyobservatory.com
Winer Observatory, www.winer.org
ADM Accessories, http://admaccessories.com/
Apogee Imaging Systems, http://www.ccd.com/
Astro Tech, https://www.astronomytechnologies.com/
Astro Hutech/BORG, http://www.sciencecenter.net/hutech/
Astro-Physics, Inc., http://www.astro-physics.com/
Atik Cameras, http://www.atik-usa.com/
Celestron, http://www.celestron.com/

Daystar Filters, http://www.daystarfilters.com/
Denkmeier Optical, Inc., http://www.deepskybinoviewer.com/
Explora-Dome (Poly Tank), http://www.exploradome.us/
Explore Scientific LLC, http://www.explorescientific.com/
Finger Lakes Instrumentation LLC, http://www.fli-cam.com/
Hotech Corp, http://www.hotechusa.com/
Howie Glatter's Laser Collimators, http://www.collimator.com/
Innovations Foresight LLC, http://www.innovationsforesight.com/
IOptron, http://www.ioptron.com/
Lunt Solar Systems LLC, http://www.luntsolarsystems.com/
Meade Instruments, http://www.meade.com/
Moonlite Telescope Accessories, http://www.focuser.com/
Peterson Engineering Corp, http://www.petersonengineering.com/sky/index.htm
Planewave Instruments, http://www.planewaveinstruments.com/
QHYCCD (Astro Factors/Deep Space Products), http://www.astrofactors.com
QSI, http://qsimaging.com/
Questar, http://www.questar-corp.com/
Santa Barbara Instrument Group, http://www.sbig.com/
Shelyak Instruments, http://pagesperso-orange.fr/shelyak/en/index.html
Sky Watcher, http://www.skywatcher.com/
Software Bisque, http://www.bisque.com/
Starlight Instruments LLC, http://www.starlightinstruments.com/
Starlight Xpress Ltd., http://www.sxccd.com/
Stellarvue, http://www.stellarvue.com/
Tele Vue Optics, http://www.televue.com/
Texas Nautical/Takahashi America, http://www.takahashiamerica.com/
Vernonscope, http://www.vernonscope.com/
Vixen Optics, http://www.vixenoptics.com/
William Optics, http://www.williamoptics.com/

Part III

Remote Observing Project Case Studies

Chapter 10

Photometry Projects

Photometry is the science of measuring light. Everything we know about stars, nebulae, and galaxies we have learned from collecting and measuring the light we receive from them. The two primary forms of information we derive from the light of these objects are brightness (magnitude) and color (frequency). A star can appear bright because it is relatively close or because it is intrinsically bright. If you know a star's distance (from measuring the parallax of nearby stars closer than 400 light years away or by its color), you can determine its absolute magnitude. The color of a star can tell you how massive it is, how hot it is, and what point it is in its evolution.

Astronomers use standardized filters for photometry measurements. These filters pass light only at a specific range of color (frequency) and block out all other light. The discrete passbands of the colors of the filters used for photometry—ranging from highest to lowest frequency—are ultraviolet (U), blue (B), visual [green] (V), red (R), infrared (I).

This chapter presents real-world case studies of photometry projects done by people using remote observatories to image objects for making high-precision photometry measurements. The figures are courtesy of the respective contributor.

Remote Variable Star Photometry Project Using the Sierra Stars 24-in. Telescope (Kevin Paxson)

I am a member of the American Association of Variable Star Observers (AAVSO) and have been observing variable stars since September 2001. Up through fall 2008, all of my observing was done visually. My primary focus was semi-regulars,

© Springer International Publishing Switzerland 2015
G.R. Hubbell et al., *Remote Observatories for Amateur Astronomers*, The Patrick
Moore Practical Astronomy Series, DOI 10.1007/978-3-319-21906-6_10

Miras, RV Tauris, and other common variable star types. Because of the poor seeing and light pollution in Spring, Texas, and more recently in Centerville, Ohio, I typically could only get down to magnitude 10.5 with my 4-in. refractor and magnitude 11.5 with my 6-in. refractor.

I became interested in remote charge-coupled device (CCD) imaging to monitor cataclysmic variable (CV) star outbursts. I wanted to "go deeper" in magnitude and make more accurate variable star estimates. Since November 2008, I have been using the Sierra Stars Observatory Network (SSON) 24-in. telescope in California, and it remains my primary choice for remote imaging and CCD photometry. In the past, I also have used Global Rent a Scope (GRAS) (and now iTelescopes.net) and the Bradford Remote Telescope (BRT) for some of my remote imaging needs.

Objectives and Goals of Project

CV Goals

My CV program goals were to monitor for CV outbursts and to determine an outburst cycle or frequency, to obtain high-quality photometry during the outburst and into quiescence, to provide greater observational coverage for many select CVs and related objects, and to announce CV outbursts to the global community, either via the Yahoo Outburst Group (https://groups.yahoo.com/neo/groups/cvnet-outburst/info) or CVnet (https://sites.google.com/site/aavsocvsection).

CVs and related objects are very closely spaced binary systems that consist of a white dwarf and a larger red dwarf. The distances involved are less than the diameter of our Sun, and the orbital periods arc only several to many hours. The red dwarf is tidally distorted and loses some of its mass over time to the white dwarf owing to gravity. Hydrogen gas from the red dwarf spirals toward and around the white dwarf, forming an accretion disc. When a critical mass is achieved, a tremendous amount of energy is released within the accretion disc, resulting in a rapid increase in stellar brightness (often up to 5 or greater magnitudes) over a period of several days. The outburst may persist for several weeks.

My cataclysmic photometry program includes some legacy CVs, a few UGWZ variable stars, some recurrent novae, and other exotic objects. Many of my program objects are from the British Astronomical Society (BAA) Variable Star Section (VSS) Recurrent Objects Program (ROP). Check out the ROP at: http://www.garypoyner.pwp.blueyonder.co.uk/rop.html.

Mira Goals

The goals of my Mira observing program were to monitor Mira variable stars through their minimum and back to maximum when able, to obtain high-quality photometry compared with the more plentiful visual observations, and to provide greater observational coverage for the less observed Miras.

Mira variable stars are red giant stars that pulsate radially, expanding and contracting owing to the "tug of war" between thermal energy and gravity after their cores have begun to fuse helium. They have periods greater than 100 days (commonly 150–450 days), with amplitude ranges greater than 2.5 magnitudes. These stars range from 0.6 to several solar masses and have diameters that extend well beyond 1.0 AU. Mira variable stars are brightest when they are at their highest temperature and minimum radius. Conversely, they are at minimum light when they are fully expanded and at their lowest temperature.

My Mira program consists of about 24 individual stars. Most of these were either under-observed or have poor CCD coverage. I try to image each program star once every 7–10 days. This offers flexibility to handle potential weather issues at the remote site.

Results

CV Results

Since I have been monitoring CVs with remote telescopes, I have detected and announced about 25–30 CV outbursts in slightly more than 6 years. It should be noted that many of the cataclysmic stars that I follow have very rare or infrequent outbursts, and I have reduced my CV observing over the last several years. I have,

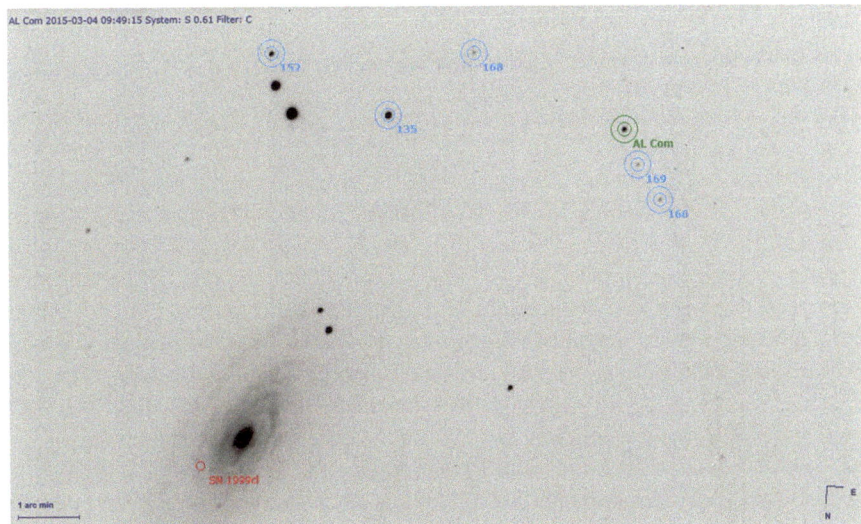

Fig. 10.1 UGWZ (UGSS?) cataclysmic variable AL Coma Berenices near galaxy M88 shown at the beginning of an outburst at 15.149 CV on 2015 March 04.40926 UT with a 30-s unfiltered exposure using the SSON 24-in. telescope. Shown on a VPhot field plot with comparison stars. With permission of the AAVSO

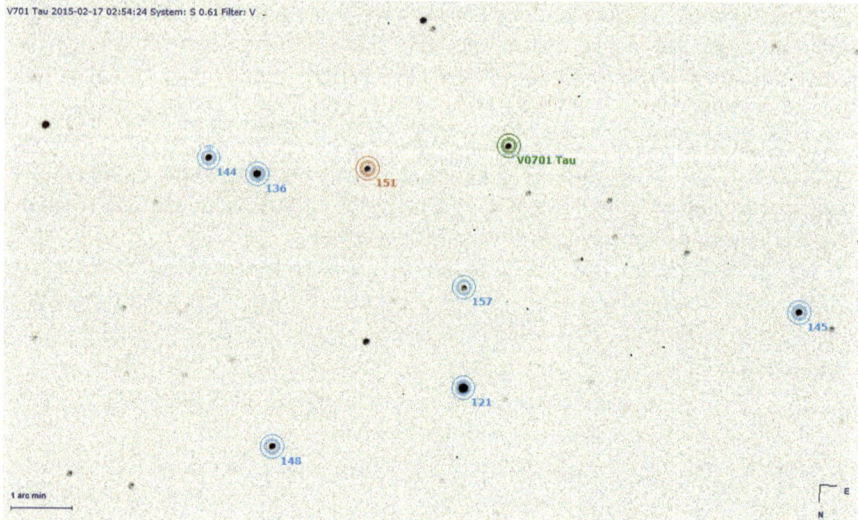

Fig. 10.2 UGSU cataclysmic star V701 Tauri shown in outburst at 14.730 V on 2015 February 18.12168 UT with a 30-s exposure using the SSON 24-in. telescope. Shown on a VPhot field plot with comparison stars. With permission of the AAVSO

however, followed many CVs after outburst to quiescence and have reported results to the AAVSO. Two examples of CV outbursts are shown in Figs. 10.1 and 10.2.

Figure 10.1 shows an outburst of the UGWZ cataclysmic star AL Com from March 2015. During outburst, this star has been observed as bright as 13.0 V and as faint as magnitude 20.0 V at quiescence. UGWZ CVs are known for their rare and infrequent outbursts (from 10 to 30 years apart). AL Com had frequent outbursts that occurred in the 1970s and 1980s and less frequent outbursts since that time, indicating a possible UGSS classification, which is given by the ROP of the BAA VSS.

An outburst image of UGSU CV V701 Tauri from February 2015 is shown in Fig. 10.2. This star can be as bright as magnitude 14.1 V during a super outburst and has a quiescence magnitude fainter than 21.1 V. UGSU CVs may show normal dwarf nova outbursts and super outbursts. Some of the SU UMa stars with long outburst intervals show interesting re-brightenings and "super humps" or light-curve modulations on their way back to quiescence after a super outburst.

It takes a lot of dedication to follow a program of CVs. As many as 97 % of your observations may be outburst non-detections because many CVs have outbursts frequencies from every 30 days to many years. This can result in many "fainter than" observations. Outburst detection often depends on how limited your observing program is, your observation cadence, clear sky availability, and sometimes luck. Consequently, I have begun to augment my CCD observing program with Mira variable stars and R CrB stars for more "quantitative" data.

Mira Results

The AAVSO is slowly moving away from visual observations and toward more CCD photometry. For most Miras, the relationship between visual estimates and Johnson V CCD photometry is not precisely known, and thus Mira CCD photometry will be very useful. The results described here for Mira star S Geminorum are typical for stars in my Mira program.

Shown in Fig. 10.3 is the Mira variable S Geminorum and its comparison stars. This Mira has an average magnitude range of 8.0–14.7 V and a period of 291 days. Figure 10.4 shows the AAVSO light curve from late August 2011 through June 2015. The Johnson V magnitudes from my SSON 24-in. telescope CCD images (shown as green circles) and visual estimates (shown as black circles) are plotted against time to yield a light curve. My observations are denoted with the green circle with blue crosshairs and were reduced using VPHOT, an online photometric reduction software package that is available only to AAVSO members.

My observations fit the AAVSO light curve very well, with good definition of the Mira troughs. There is some scatter in the visual data and some of the other CCD data. It is common for visual observations to be fainter at Mira cycle lows and brighter at Mira cycle peaks because of the human eye's sensitivity to red light. Overall, I am very satisfied with my Mira program results.

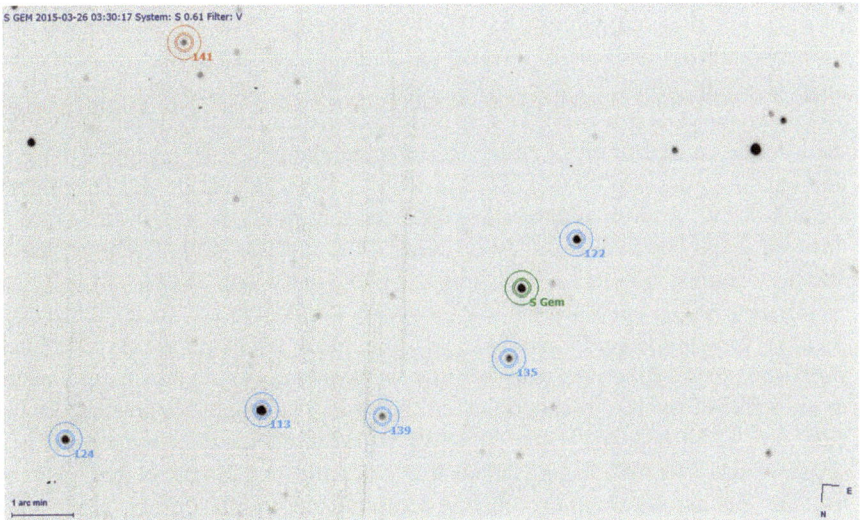

Fig 10.3 A VPhot field plot showing S Geminorum at magnitude 12.091 V on 2015 March 26.14603 UT (from a 30-s exposure with the SSON 24-in. telescope) and the sequence of its comparison stars. With permission of the AAVSO

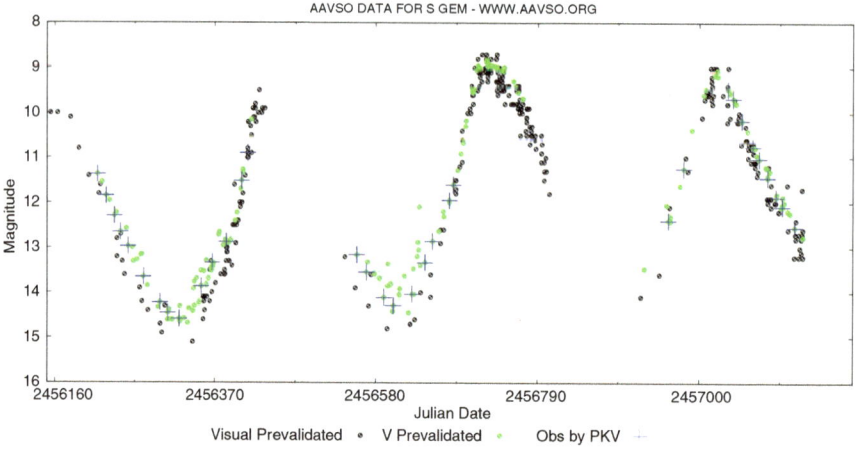

Fig 10.4 A light curve of Mira variable S Geminorum from late August 2011 through June 2015 with observations by the author (*green circles* and *blue crosshairs*). With permission from the AAVSO

Benefit Derived from Using Remote Observatory Facilities

I find remote imaging with the SSON 24-in. telescope to be a great value. If you were to minimally replicate its imaging capability, you would have to invest a minimum of $20,000 for a Paramount ME mount, a 14-in. Schmidt Cassegrain OTA, and a top-end CCD camera. Taking 90-s exposures with this setup would be within the imaging capability of the SSON 24-in. telescope with a 30-s exposure. I find this initial startup cost to be high, and with my low percentage of clear nights, I could do 20 years of remote imaging at current prices before I would match this initial $20,000 capital outlay of doing CCD imaging myself. This is major reason I continue to do remote imaging and photometry with SSON.

I find the SSON 24-in. telescope to be very convenient to use. Image scheduling is easy with a user-friendly interface, and my images are automatically calibrated with darks, flats, and artifact maps before being available for .ftp transfer or transfer to my AAVSO VPHOT account. One of the major challenges for amateurs doing their own CCD imaging is the taking of flats and accurate calibration of their images. I schedule my imaging targets in the evening, and usually before noon the next day, my images are in my VPHOT account ready for photometric reduction.

I also appreciate the light collection capability of a large remote telescope. When using the SSON 24-in. telescope with 30-s unfiltered exposures, I can typically get down to magnitude 18 with an accuracy of ±0.2 magnitudes. With a 30-s image with a Johnson V filter, I can get down to about magnitude 15.5 with an accuracy of ±0.05 magnitudes. CCD pixel saturation occurs at about magnitude 10.7 with a Johnson V filter with a 30-s exposure, and a 10-s Johnson V exposure saturates around magnitude 8.5–9.0, depending on sky conditions.

Because CVs are typically uncolored during quiescence and outburst, using unfiltered CCD exposures for outburst detection is fully acceptable. However, once a CV outburst has been detected, filtered photometry is recommended. A Johnson V filter also is required for Miras and other variable stars that exhibit color. You could also take B-filtered images and shoot "standard fields," such as M67, to "transform" your photometric results for even more accurate data.

Stellar photometry using remote imaging is a great way to do real science if you don't want to make a large capital investment. If you are interested in variable stars, variable star observing, or CCD photometry, check out the AAVSO website at: www.aavso.org. To learn more about CVs, visit the CVnet website at: https://sites. google.com/site/aavsocvsection. To learn more about Mira and other long-period variable (LPV) stars, check out the AAVSO LPV website at: https://sites.google. com/site/aavsolpvsection/Home.

Equipment Used

I used the SSON 24-in. (0.8-m) telescope in California for the work I describe in this case study. The details of the instruments used are:

- Telescope: 24-in. (0.8-m) F/10 Classical Cassegrain
- CCD camera: FLI PL09000 CCD camera with 12-μm pixels, a 21 by 21 arc-minute field of view, and a peak quantum efficiency of 60 %
- Filter used: Standard Johnson-Cousins BVRI photometric filters

About Kevin Paxson

Kevin B. Paxson is an amateur astronomer who has been a member of the American Association of Variable Star Observers since 2001. He served on the AAVSO Council (Board of Directors) from 2012 to 2014. Kevin is now retired after working nearly 30 years in the petroleum industry in New Orleans and Houston. He now resides in Centerville, OH.

Exploring the Physics of Binary Systems by Looking at Period Variation (George Faillance, Carl Owen, David Pulley, Derek Smith, Americo Watkins)

Our group consists of five friends with a common interest in astronomy and astrophysics, four of whom met while working on an Open University project. Our fifth member joined through our membership in the British Astronomical Association (BAA).

We were introduced to the idea that the physical processes taking place in stars and between the stars in a binary system can be glimpsed by measuring subtle changes in the period and light-curve of the system. These processes can have a number of potential causes, but each has a particular effect on the pattern of change in the period. To establish these patterns means observing many cycles over long periods of time. From an amateur astrophysicist's point of view, the shorter the period the better because you can observe more cycles in a given time. Subdwarf B type (sdB) stars have some of the shortest periods observed in binary systems—typically 2–3 h.

HS 0705+6700 is an example of such a system; it has a period of 2.3 h. Therefore, it is an ideal study for an amateur research group such as ours. HS0705 is a member of the HW Vir family of short-period binary systems that consist of an sdB star and a cool, low-mass, main-sequence star or brown dwarf. Their compact structure and the large temperature difference between the two components give rise to short and well-defined primary eclipses, allowing times of minima to be determined with high precision. The sdB components of these systems have typical masses of about $0.5~M_\odot$ and consist of a helium burning core with a thin hydrogen envelope. They are located at the left-hand extremity of the horizontal branch in the H-R diagram.

We were introduced to sdB type systems through work we performed for the Keele University Department of Physics and Astrophysics in the UK on the binary system J1628+10. This work involved the use of the Open University's Mallorca-based PIRATE remotely controlled telescope, normally reserved for students. This work introduced us to the world of sdBs.

SdBs are systems whose primary star has evolved to helium burning and that also have a small main sequence secondary. During this evolution, a common envelope developed that drained angular momentum from the system and caused the orbit to shrink to the point where periods are much shorter than normal. A consequence of this evolutionary path is that they tend to have extremely hot primaries and very cool secondaries. In addition, some have been found to have circumbinary companions (exoplanets, brown dwarfs, etc.). This fact is a subject of debate among our professional colleagues because of the possibility that these circumbinary objects evolved in the conventional way planets evolve during the process of star system evolution, akin to our Solar System, or that they are a product of and a contributor to the collapse of the common envelope—maybe a combination of the two. The presence of circumbinary objects is indicated by cyclical type variations in the eclipse timings. (An example appears later in Fig. 10.6.) However, other effects, such as apsidal motion or gravitational coupling of the binary orbit to variations in shape of magnetically active stars (the Applegate effect), can produce similar effects. For HS0705, both of these effects are often considered too small to influence the eclipse timings, and the presence of the circumbinary object is the preferred explanation.

Objective and Goal of Project

During the process of identifying HS0705+6700 as an object to study, we reviewed some 83 eclipsing binary systems. Some were rejected because of their visibility from the northern hemisphere, some were too faint, or had too long a period, insufficient historic data, too complex, or possibly already well researched. Having eliminated about half of the candidates on this basis, we voted, and HS0705+6700 was chosen.

Our approach was to break the research into two areas:

- Confirm the period of the system and identify any variations in that period over time.
- Generate a light-curve and from it infer some of the system parameters.

We did not have access to spectrographic data, other than that published by other researchers.

It quickly became clear that we were more successful in analyzing the period and its variation over time than we were in analyzing the light-curve. In time, it became clearer where our problems lay. These systems have high surface gravities, with the primary being extremely hot and luminous; this, coupled with the close proximity of a small M dwarf secondary, generates complex reflection effects that we, like our professional colleagues, have difficulty modeling. This learning curve is still being climbed.

We were much more successful in analyzing and interpreting the variations in period over time. We made this the primary goal of our efforts. A paper presenting the results of our research was accepted for publication by the BAA (to be published shortly), which also provided part funding for this research, for which we are very grateful.

Results

HS0705+6700 provides a deep, 1 magnitude, well-defined primary eclipse as would be expected from a system with an orbital inclination of 840 with a cool secondary M dwarf star having a diameter just 20 % less than its hot subdwarf companion. At 0.15 magnitude, the secondary eclipse is considerably less well defined. Figure 10.5 shows the very clean primary minimum obtained in March 2015 with the Sierra Stars Observatory Network (SSON) 24-in. F/10 Optical Mechanics Nighthawk CC06 telescope with a V filter. The sloping shoulder leading into the eclipse is attributed to the decreasing amount of reflected light from the cool secondary star as the eclipse progresses toward totality.

With both stars of HS0705+6700 lying within their respective Roche lobes, there can be no mass transfer in this detached system. The orbital rotation can be likened to the pendulum of a precision clock where each primary eclipse is separated from the next by the binary system period. For HS0705+6700, the binary period is 0.095646671 days ±4 ms (2 h 17 m 43.872 s ±0.004 s); this precision allows us to make confident predictions for future eclipses. Comparing these predictions with the observed eclipse times, we can investigate the many influences exerted on this binary system. This process leads us to construct the important Observed minus Calculated (O−C) diagram—a graphical representation of eclipse timing errors (or residuals).

The calculated times of primary minima are determined from the system's ephemeris by $C = T_0 + P*E$, where C is designated as our calculated time of

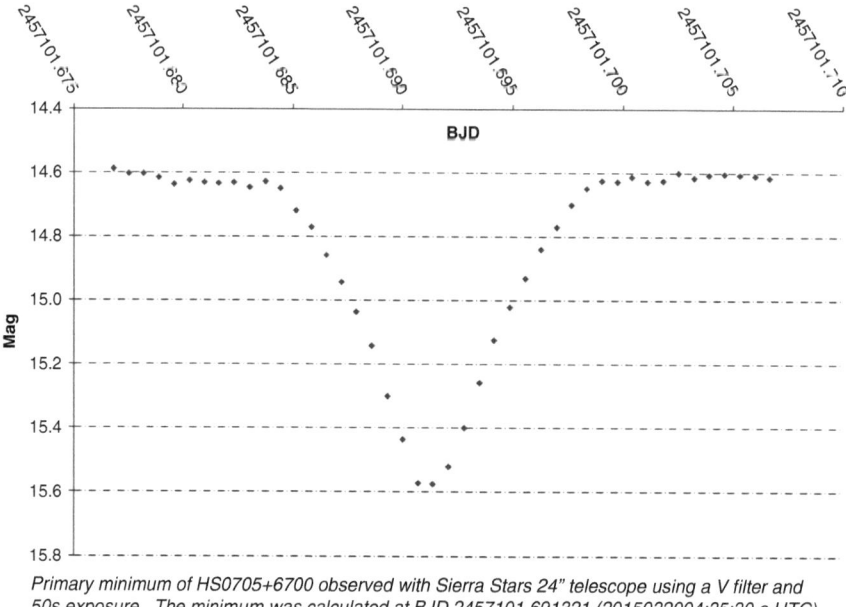

Primary minimum of HS0705+6700 observed with Sierra Stars 24" telescope using a V filter and 50s exposure. The minimum was calculated at BJD 2457101.691321 (2015032004:35:30.o UTC)

Fig. 10.5 Primary minimum of HS0705+6700 with SSON 24-in. telescope using a V filter and 50-s exposure. The minimum was calculated at BJD 2457101.691321 (20150320 04:35:30. UTC)

minima, T_0 is the time of first minimum, P is the binary period, and E is the cycle number starting from 0 and increasing in integers 1, 2, 3..., with each new eclipse.

Measurement of the position of the minimum, designated O, was made by taking a series of observations starting typically 10–15 images before the minimum and with a similar number post minimum. For HS0705+6700, using SSO's 24-in. telescope with a V filter, imaging would typically start 15 min before the minimum and end 15 min after, taking between 30 and 35 images. Exposure duration was kept as short as possible so that the resulting data points on the light-curve were kept close together, but sufficiently long to achieve an acceptable signal-to-noise (SNR) to minimize data scatter. We opted for an exposure of 50 s, giving a typical SNR ratio of greater than 120 for this 15th magnitude object. Photometry of the resulting FITS images was carried out using the proprietary software MaxIm DL® but other software can be used, e.g., AIP4WIN, resulting in a light-curve similar to Fig. 10.5.

The position of the minimum was calculated using the Kwee and van Woerden methodology encoded in the proprietary software package Peranso®. The free software Minima25c (Bob Nelson) was also used to confirm consistency in the computed results. Other geometric and computational methods are available but the Kwee and van Woerden methodology is widely used and, when combining results with those from other investigators using the same methodology, minimizes the risk of introducing offsets that can occur with the use of mixed approaches.

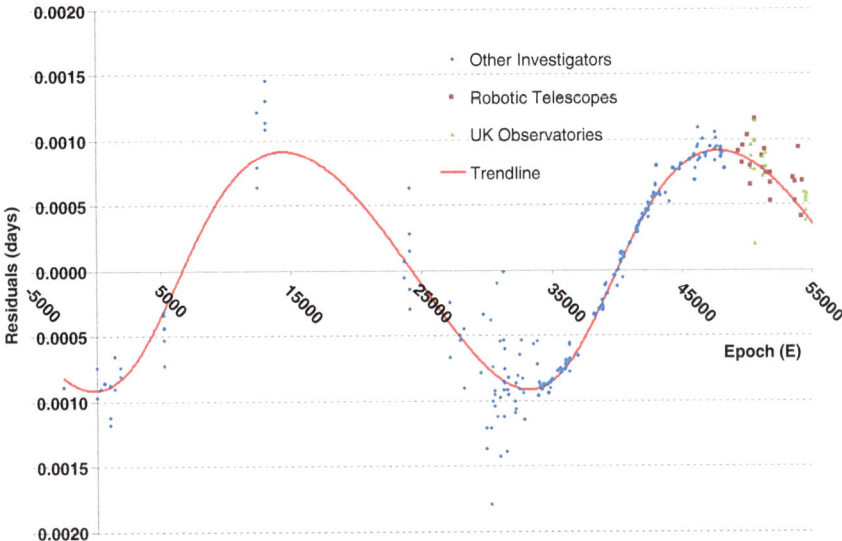

The (O-C) residuals for HS0705+6700 from February 2000 until March 2015. The red curve represents the best fit for a circumbinary object in an elliptical orbit around the binary pair. The circumbinary object and the binary pair orbit the system's barycentre. It is the motion of the binary pair around the barycente that causes the (O-C) variations and is often referred to as Light Travel Time Effects.

Fig. 10.6 The (O−C) residuals for HS0705+6700 from February 2000 until March 2015. The *red curve* represents the best fit for a circumbinary object in an elliptical orbit around the binary pair

Between September 5, 2013, and March 26, 2015, we obtained 42 instances of minima of HS0705+6700 using SSON and iTelescope's robotic telescopes and the UK observatories of group members George Faillace and Americo Watkins. We have added our 42 data points to the 189 known historical data points taken by Drechsel, Niarchos, Qian, Camurdan, Beuermann, and others between 2000 and 2013. From this dataset, we constructed a new ephemeris and plotted the (O−C) residuals shown in Fig. 10.6. The pattern of the residuals suggests that there is a possible cyclical influence affecting the eclipse timings.

Causes of these potential cyclical variations include (1) apsidal motion, (2) magnetic quadrupole effects, and (3) the presence of a circumbinary object(s). Apsidal motion and magnetic quadrupole effects are usually considered to have an insignificant impact on the eclipse timings of detached short-period binary systems with a cool dwarf secondary. Most investigators, particularly Qian and Beuermann, consider a circumbinary object as the most likely driver for these timing variations.

Assuming a circumbinary object is present, we have estimated the orbital parameters of a potential third body by finding the best fit sine/cosine curve to the residuals of Fig. 10.6. This fit is shown as the red curve, which suggests that if a body does exist, it would have a probable mass similar to a brown dwarf or small star orbiting at 3.8 AU and with a period of approximately 9.0 years. The circumbinary object and the binary pair orbit the system's barycentre. It is the motion of the binary pair around the barycentre that causes the $(O-C)$ variations and is often referred to as "light travel time effects."

While a number of claims of circumbinary objects orbiting binary systems have been made in recent years, many have since been disproved using techniques of orbital stability analysis and, in the case of V471 Tau, direct observation with ESO's 8.2-m VLT at Cerro Paranal. The jury remains out on the possibility of a circumbinary object around HS0705+6700.

A fuller analysis of HS0705+6700, and details of the references mentioned herein, can be found at http://arxiv.org/abs/1502.04366 which is to be published in the *BAA Journal*.

Benefit Derived from Using Remote Observatory Facilities

Our group consists of five individuals, two of whom have their own observatories. We are based in the southern part of the UK. The benefits of having access to remote facilities are manifold:

- It allows the three members who do not own telescopes to contribute to group observations.
- It allows the group to observe systems that may not be visible from the UK because of low declinations.
- It allows observations when the weather in the UK may not be favorable.
- It provides access to state-of-the-art robotic telescopes with a range of apertures, filters, and sensitivities so that the most appropriate one can be selected.

There are a number of reasons that sdB type systems are ideally suited to being observed from remote observatories:

- Because of their short periods, a whole light-curve can be observed at relatively modest cost; minima even more so.
- The minima are clearly defined.
- They are a relatively unstudied group of binaries and so, as amateurs, we can contribute something of real scientific value.

As a result of our work with remote observatory facilities, we have published three papers, one on HS 0705+6700, with follow-up papers in the pipeline.

Facility Equipment Used

Table 10.1 lists the facilities used in the production of the HS0705+6700 paper.

Table 10.1 List of facilities used

Observatory	Telescope	Instrumentation
Remotely Operated Telescopes		
Sierra Stars Observatory Markleeville, CA http://sierrastars.com/gp/ SSO/SSO-CA.aspx	0.61 m FL 6100 mm	Finger Lakes Inst. ProLine camera 3056 by 3056 pixels FOV 21 by 21 arc-minutes
Sierra Stars Observatory Mount Lemmon, AZ http://sierrastars.com/gp/ MLSC32/MLSC32.aspx	0.81 m FL 5670 mm	SBIG STX KAF-16803 camera 4096 by 4096 pixels FOV 22.5 by 22.5 arc-minutes
iTelescope T21 Mayhill, NM http://www.itelescope.net/ telescope-t21/	0.43 m FL 2920 mm	Finger Lakes Inst. PL6303E camera 3072 by 2048 pixels FOV 49.2 by 32.8 arc-minutes
Group Member Owned Telescopes		
Woodstoke Observatory Oxfordshire, England	0.28 m FL 1400 mm Celestron	SBIG ST-9XE camera 512 by 512 pixels 20 by 20 μm pixel size 10.2 mm × 10.2 mm array FOV 25.1 by 25.1 arc-minutes
Astrognosis Observatory	0.356 m FL 2736 mm Celestron	QSI 532E camera 2184 by 2184 pixels FOV 18.7 by 12.6 in.

About George Faillace, Carl Owen, David Pulley, Derek Smith, and Americo Watkins

George Faillace holds bachelor's and master's degrees in Civil Engineering and an MBA. He worked in engineering consultancy and in an international development institution in the United States, and as independent consultant in the UK until retirement. At The Open University in the UK, George studied astronomy and mathematics, gaining a degree in 2012. Astronomy has been his hobby (and passion) from a very early age.

Carl Owen holds a BSc (Hons) in Natural Science. He works in the warehouse and distribution industry, where he began his career after leaving college. Studying with The Open University alongside full time work, Carl fed his hunger for knowledge and through project work, was introduced to remote telescopes.

David Pulley holds a degree in Electronic Engineering and spent his early years conducting industrial research programs into microwave semiconductors with applications in both the telecommunications and radar sectors. He then moved into manufacturing management where he was Operations Manager of the electronics production facility of a UK multinational company supplying vacuum equipment to the scientific and semiconductor industries. David became the company's Global Quality Manager before taking early retirement in 2007 to pursue his interests in mathematics, physics, and astronomy.

Derek Smith holds a BSc (Hons) in Physics and an MBA. He worked in broadcasting/theatre from university until retirement, initially as an engineer and then in various technical management roles. Derek took astrophysical courses with The Open University in the UK preparatory to a PhD in retirement. However, he met up with this group and decided to pursue research with this group instead.

Americo (Eric) Watkins holds a bachelor's degree in Physics and an MSc in Astronomy. He has been actively interested in astronomy since the age of 12. From university, Eric joined the police force. On retirement from the police force, he engaged in further postgraduate study in astronomy. Eric now pursues his research interest in eclipsing binaries with this group. Other research interests are active asteroids and centaurs.

Comparison of Visual and CCD Magnitudes of Comets (Roger Dymock)

A significant advance in amateur astronomy in recent years has been the setting up and use of remotely operated or robotic telescopes. Use of such facilities is not expensive, especially when compared with the cost of setting up one's own observatory from scratch. All of the sky, not just the hemisphere in which you are located, is available to robotic telescope users and not necessarily in the middle of a cold night. Unfortunately, UK-based observers are often at a disadvantage because of poor weather, light pollution in towns and cities, or the fact that our latitude (50–56 %), prevents access to many celestial phenomena further south, and also means that our summer nights are very short.

A proposal to establish a new position—Robotic Telescope Coordinator—and to set aside funds to promote an active program of remote observation by BAA members—was put to the BAA Council on 2008 April 30, 2008, by Dr. Richard Miles and Roger Dymock. It received unanimous support, and the BAA's Robotic Telescope Project was born. To participate, observers submit a proposal to the coordinator, which is then reviewed by experienced BAA members. If the proposal is accepted, 50 % of the estimated credits required are added to the observer's Sierra Stars Observatory Network (SSON) account.

In January 2013, a project "to prove the proposed Comet Section methodology for extracting visual equivalent or total magnitudes from charge-coupled device (CCD) images" was submitted to and approved by the BAA. That methodology has proved successful and is the subject of this section. Subsequently, very bright (greater than magnitude 8) comets proved to be a problem in that the CCD magnitudes tended to be fainter than the equivalent visual measures. The use of Iris, described later, has overcome this difficulty.

CCD Astrometry and Photometry

Previous to the implementation of this project, visual comet observers have generally estimated comets to be brighter than CCD imagers have. The reason is that visual estimates (total magnitude or m1) include the whole of the comet whereas CCD measures relate to the central or nuclear portion only (nuclear magnitude or m2). The methodology is summarized here and the procedure is described in detail on my Project Alcock website (http://www.britastro.org/projectalcock/index.htm) under the heading "CCD Astrometry and Photometry."

Generating visually equivalent magnitudes (plus coma diameters, tail lengths, and position angles) from CCD images enables:

* Integration of CCD observations with visual observations
* Production of light-curves for a greater part of a comet's orbit (when the comet is too faint for visual observers)

Methodology

The methodology includes the following steps:

* Measure the position and magnitude of the comet (using Astrometrica and FoCAs)
* Obtain a visual equivalent magnitude using Kphot
* Measure the coma diameter using AIP4WIN and Microsoft® Excel or any equivalent software
* Measure tail length and position angle using Astrometrica or Aladin Sky Atlas
* Report the relevant measurements to the Minor Planet Center (MPC), Comet Observation DataBase (COBS), and other national and international organizations

Images must have been obtained using a clear filter or no filter and calibrated (application of dark frames, bias frames, and flat fields). Exposure times must be such as to obtain a reasonable signal-to-noise ratio (SNR) and to avoid trailing. Figure 10.7 is an image of comet 29P/Schwassmann-Wachmann obtained on February 26, 2015, using the Warrumbungle (Australia) telescope. This particular comet is prone to cyclical outbursts and had brightened from magnitude 16.5 on January 21, 2015, to magnitude 12.6 on February 17. I measured it to be magnitude 13.1 at the time of this image.

The software packages mentioned here are running under Windows® XP— Astrometrica, FoCAs, and Kphot must be used in that order to measure magnitude and position. Measurement of coma diameter, tail length, and position angle using AIP4WIN and Microsoft® Excel, Astrometrica, or Aladin Sky Atlas can be made at any time.

Although you can follow the procedure through all its steps, you may, of course, not want to do so. The alternatives are:

1. Process images with Astrometrica and generate MPC reports from Astrometrica
2. Process images with Astrometrica and FoCAs and generate MPC and multibox reports from FoCAs

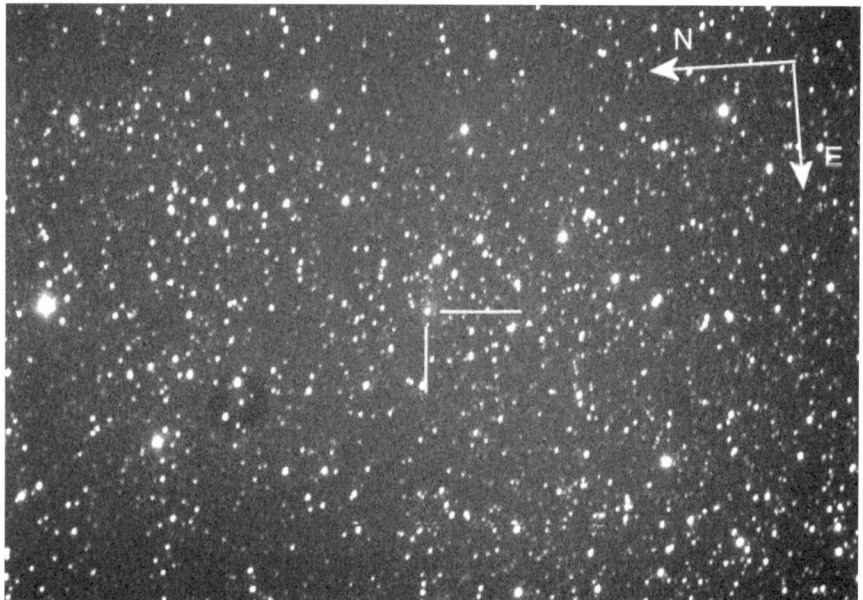

Fig. 10.7   Comet 29P/Schwassmann-Wachmann

3. Perform step 2 and also process images using Kphot plus the other applications
   mentioned above to produce a complete *International Comet Quarterly* (ICQ)
   formatted report

You will need an MPC Observatory Code before reporting comet CCD astrometry and photometry to the MPC but not necessarily for reports to other organizations. However, obtaining one would give others greater confidence in your reports.
The MPC's *Guide to Minor Body Astrometry* describes how to do this in detail, in
particular paragraphs 15 and 16. It is also covered in Chap. 10 of my book *Asteroids
and Dwarf Planets and How to Observe Them* (Springer 2010).

Astrometry

Astrometrica, which I highly recommend, is used to make astrometric measurements. Before processing images, it is advisable to ensure your comet (and asteroid) files are up to date:

- To download the latest comet elements, start up Astrometrica and go to Internet/
  Update MPCorb and select Comets
- To download the latest asteroid elements, start up Astrometrica and go to
  Internet/Download MPCorb

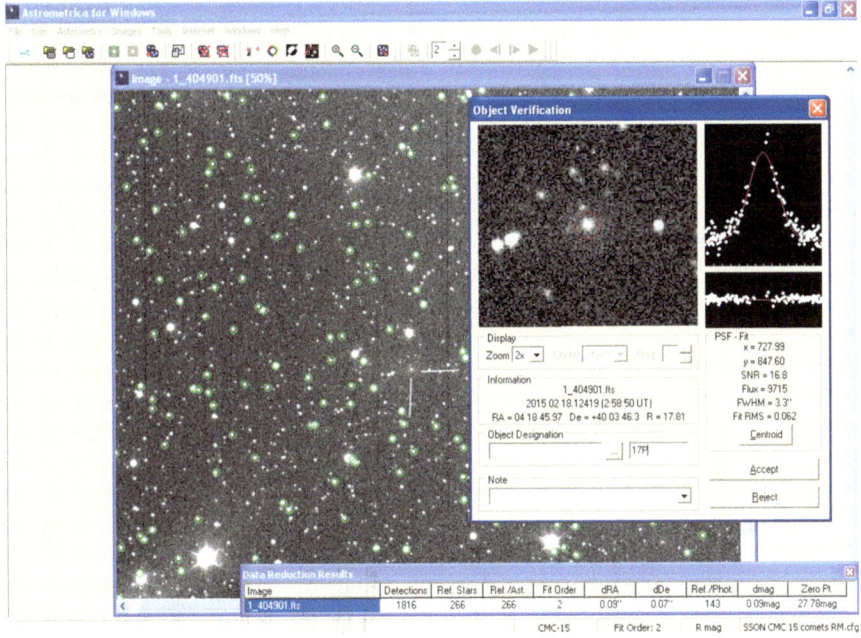

Fig. 10.8 Screenshot of Astrometrica showing image, Data Reduction Results and Object Verification Window. Comet 17P is indicated

To ensure compatibility with recognized methods of measuring comet magni-tudes, set the Aperture Radius (in pixels) in the Astrometrica configuration file to the equivalent of an actual radius of 5.5 arc-seconds.

Stacked images cannot be used because FoCAs cannot handle them. Images must be processed twice with Astrometrica. Use the UCAC 4 catalog for astrometry first and then the CMC-14 or 15 or the USNO-A2.0 catalogs for photometry sec-ond. USNO-A.2 should only be used if not enough reference stars are available using CMC-14 or 15.

For each of the two runs, open the images you want to process and click the "Astrometry" button. Place the crosshairs over the head of the comet (do not use the Control key when doing this; the Centroid option can be used) and click the left mouse button (Fig. 10.8). In the "Object Verification" window, click the but-ton under "Object Designation" and select the comet from the list. If there is a large error (dRA and/or dDe greater than say .02′), then you may have the wrong object (or your MPCorb catalog is out of date). Clicking "Accept" adds the comet ID to the image and generates an MPC report. After processing all the images, the completed MPC report can be viewed by selecting "File/View MPC Report File."

Photometry

The Spanish Group's (Cometas) FoCAs software processes the Astrometrica output to calculate magnitudes for a range of apertures (multiboxes). Having completed the Astrometrica processing, start up FoCAs. If asked, delete any previous MPC and multibox data. Clicking Images brings up a screen that lists the number of images processed with Astrometrica with both catalogs. Highlight the images you want to process and then select "Process" to display the results. MPC and Multibox reports can be saved and then emailed to the previously nominated recipients.

Visual equivalent magnitude. Kphot, written by Uwe Pilz, generates visual equivalent magnitudes, m1—what a visual observer would see and measure—from FoCAs output, which must be saved as kp.txt in the Kphot folder.

In addition, NET software needs to be installed and a comets data file downloaded from the MPC. Note that the comet data file must be renamed from Soft02Cmt.txt to comets.dat. There are a number of comet files in different formats on the MPC website so make sure you download the correct one. To run Kphot, open a Command Prompt window and change the directory to the Kphot folder.

Coma diameter. Any software that will allow you to produce a profile and export the data to a spreadsheet will suffice, but I use AIP4WIN and Microsoft® Excel.

To produce the profile, start AIP4WIN, open an image, and select "Measure/ Profile" tool. Draw a line through the head of the comet (not including the tail) and save the data by selecting "Save Profile in Data Log."

Import the data into Microsoft® Excel, plot a graph of ADUs versus pixels, and count the number of pixels between the points where the curve merges with the background level. The diameter of the coma can then be obtained by multiplying the number of pixels by the size of the pixels in arc-seconds.

Tail length and position angle. The Project Alcock website describes three ways to measure tail length and position angle but, for the sake of brevity, only one is mentioned here. Open an image in Astrometrica and astrometrically calibrate it (depending on your imaging software, the image may already be in that condition—if RA and Dec values are shown at the bottom of the screen as you move your mouse across the image). While holding the shift key down, move the mouse from the head of the comet to the tail; the length and position angle will be displayed below the image.

Photometry of very bright comets. As mentioned earlier, the procedure described above does not cope well with very bright (greater than magnitude 8) comets but Iris does give reasonable results. Before measuring the magnitude of the comet with Iris, the Magnitude constant (Iris terminology for Zero Point) must be established using Astrometrica. Having done that, the coma diameter is measured using the Iris Slice function (Fig. 10.9). Knowing both the Magnitude Constant and coma diameter the magnitude of the comet can be obtained using Iris Aperture Photometry. You could, of course, go outside with binoculars or a telescope and make a visual estimate!

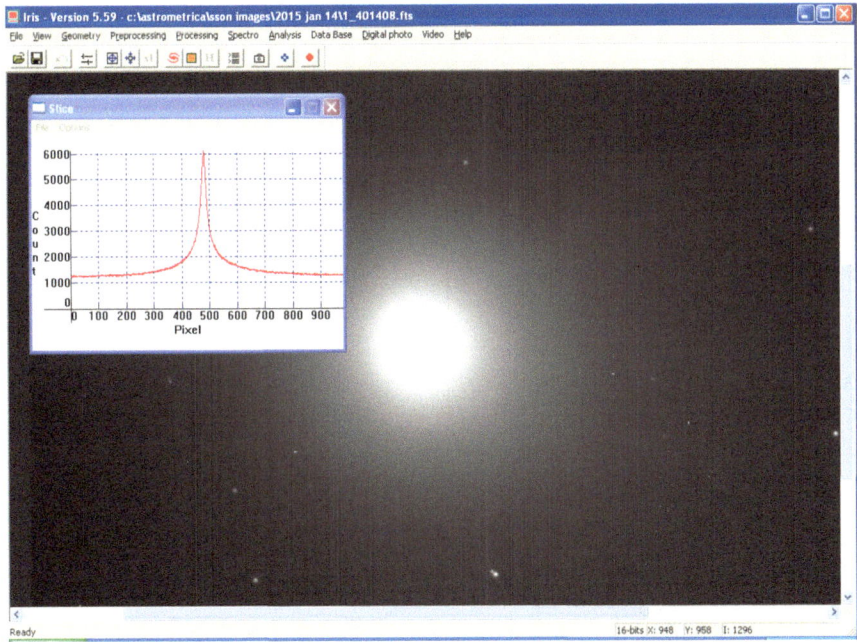

Fig. 10.9 Iris Slice facility used to measure coma diameter of bright comet C/2014 (Lovejoy) imaged with the SSON telescope January 14, 2015

ICQ format reporting. By working through the total procedure you now have all the data necessary to file a report in ICQ format on the COBS database. Once the various ICQ codes have been negotiated and a set of observations for your telescope/CCD combination has been entered, adding further observations is much simpler. Note that the COBS database may use codes not referred to in either of the ICQ code listings but if this is the case, then use the COBS codes.

Data for selected comets over a specified time period can be displayed, as can your own observations for all or a specific comet. Using this facility, observations can be downloaded and sent to your preferred organizations.

Results

The light-curve of comet C/2012 X1 (LINEAR) (Fig.10.10) shows how well visual and CCD magnitude measurements can be integrated using the method described here. The ability of CCD observations to extend the light-curve to cover several months is also illustrated. The light-curve includes magnitude estimates I made using the SSON telescope and those made by BAA member Kevin Hills using his own robotic telescope plus other CCD and visual observers.

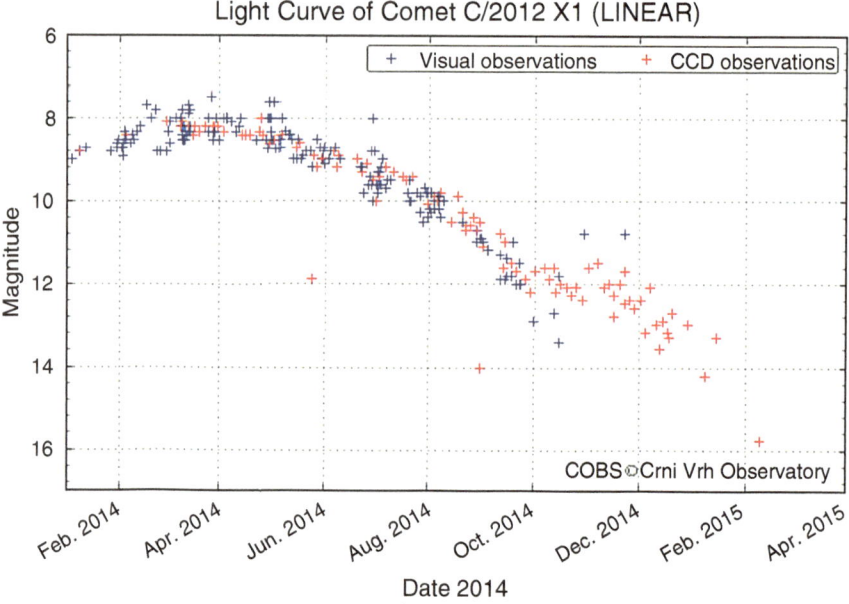

Fig. 10.10 Light-curve showing compatibility of visual and CCD magnitude estimates. Credit COBS

Figure 10.11 shows the light-curve for comet 29P/Schwassmann-Wachmann, which is of particular interest because it is prone to regular outbursts. I was only able to image it by using the SSON Warrumbungle robotic telescope located in Australia. Visual equivalent CCD magnitude measurements have allowed extension of the light-curve to fainter magnitudes than obtainable by visual measurements alone.

Benefit Derived from Using Remote Observatory Facilities

The following benefits can be derived from using remote observatory facilities:

- Large aperture telescope, which means that fainter and faster moving comets can be imaged
- Wide range of filters available
- Northern and southern hemisphere telescopes, which allow comets to be imaged that are not usually visible to observers located in the opposite hemispheres
- Much more comfortable than being outside at all hours on cold nights
- Better observing conditions, including minimal light pollution and seeing generally better than observing from or close to urban areas

Remote facilities also benefit observers with a disability that prevents them from using their own telescopes or those located in heavily light-polluted urban areas. Observers whose work means they are always on the move can use such facilities wherever they are in the world, provided they have an Internet connection.

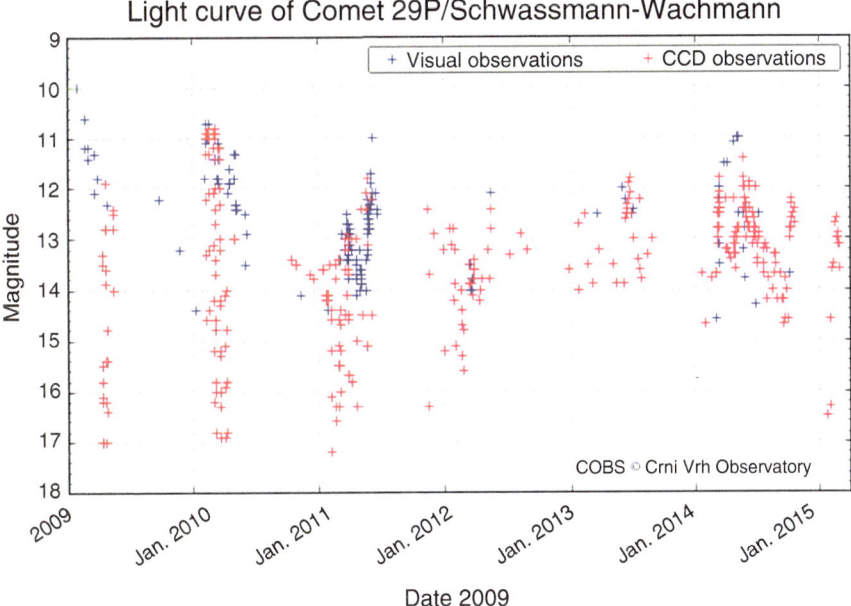

Fig. 10.11 Light-curve showing variability in magnitude of comet 29P. Credit COBS

Facility and Equipment Used

I have used both the SSON telescope in California and the Warrumbungle telescope in Australia (and very occasionally the Rigel and Mount Lemmon telescopes in Arizona), mainly unfiltered but occasionally using an R filter, for comet imaging. A venture into spectroscopy has made use of the SSON transmission grating spectrograph.

About Roger Dymock

Roger Dymock lives in Waterlooville in Hampshire on the south coast of England. In recent years, Roger has become—as he describes himself—somewhat of a fair-weather observer—late, cold nights are not for him. Most of his imaging is done with the SSON telescope, and he is currently experimenting with iTelescope. Roger also owns a 10-in. Orion Optics (UK) Newtonian reflector housed in a pod in his back garden. A WiFi link enables him to monitor it from his study once it is set up and running.

Roger's interest in astronomy did not become serious until he took early retirement in 1996. A member of the BAA and RAS for many years, he also belongs to the Hampshire Astronomical Group, which has an observatory with several telescopes located in Clanfield, Hampshire.

During his time with the BAA, Roger has served on Council and been Asteroid and Remote Planets Section Director. After writing a book—*Asteroids and Remote Planets and How to Observe Them* (Springer 2010)—he decided it was time for a change. Currently, he is an active member of the Comet Section with a mentoring and outreach role. To better fulfill that role, he pulled together several existing pieces of software to allow visual equivalent magnitudes to be calculated from CCD images. This procedure and others to encourage observers to develop their comet imaging skills can be found on Roger's Project Alcock website at http://www.britastro.org/projectalcock/index.htm. Recently, he has experimented with spectroscopy again using the SSON telescope. Unfortunately, no really bright comets have appeared of late, but imaging stars has allowed him to gain experience in both observing and image processing.

References

EL CVn-type binaries-Discovery of 17 Helium white dwarf precursors in bright eclipsing binary systems Mon. Not. R. Astron. Soc. 000, 1–12 (2009) Printed 24 September 2013. Also: arXiv:1310.4863

Pulley D et al. The eclipsing binary HS0705+6700 and the search for circumbinary objects, arXiv:1502.04366

British Astronomical Association (BAA), https://britastro.org/

BAA Robotic Telescope Project, http://www.britastro.org/robotscope/

Sierra Stars Observatory Network, http://www.sierrastars.com/Default.aspx

Project Alcock, http://www.britastro.org/projectalcock/index.htm

CCD Astrometry and Photometry, http://www.britastro.org/projectalcock/CCD Astrometry and Photometry.htm

Astrometrica, http://www.astrometrica.at/

FoCAs, http://www.astrosurf.com/orodeno/focas/

Kphot, downloadable from http://kometen.fg-vds.de/kphotsoftware.zip

AIP4WIN, http://www.willbell.com/aip/index.htm

Microsoft® Excel, http://www.microsoftstore.com/store/msusa/en_US/pdp/Excel-2013/productID.259321300

Aladin Sky Atlas, http://aladin.u-strasbg.fr/

Minor Planet Center (MPC), http://minorplanetcenter.net/

Comet Observation Database (COBS), http://www.cobs.si/

MPC's *Guide to Minor Body Astrometry*, http://www.minorplanetcenter.net/iau/info/Astrometry.html

Dymock R (2010) *Asteroids and Dwarf Planets and How to Observe Them*, http://www.amazon.com/Asteroids-Planets-Observe-Astronomers-Observing/dp/144196438X

Spanish Group (Cometas), http://www.astrosurf.com/cometas-obs/

Kphot, can be downloaded from http://kometen.fg-vds.de/kphotsoftware.zip

NET software, http://www.microsoft.com/net

Comets.dat file from the MPC, http://www.minorplanetcenter.net/iau/Ephemerides/Comets/Soft02Cmt.txt

Iris, http://www.astrosurf.com/buil/us/iris/iris.htm

International Comet Quarterly (ICQ), http://www.icq.eps.harvard.edu/

ICQ Keys to Codes, http://www.icq.eps.harvard.edu/ICQKeys.html

ICQ Recommended sources for stellar magnitudes, http://www.icq.eps.harvard.edu/ICQRec.html

Chapter 11

Astrometry Projects

Astrometry is the science of measuring the precise position of objects in space. Everything in the universe is moving. Some things, however, move quicker than others from our perspective on Earth. Nearby stars move perceptibly against the background of more distant stars over a span of years owing to the phenomenon called parallax, while distant stars and galaxies don't appear to move perceptibly over the span of a human lifetime. Solar system objects on the other hand are in constant motion against the star background as they orbit the Sun. To determine the precise orbit of a solar system object such as an asteroid or comet, astronomers need to measure its precise position over a period of time. The more measurements acquired along the arc of the orbit, the more accurate the orbit determination becomes. An astrometry measurement is composed of three parts: the right ascension, the declination coordinates, and the exact time the coordinates were measured.

This chapter presents real-world case studies of astrometry projects done by people using remote observatories. The figures are courtesy of the respective contributor.

© Springer International Publishing Switzerland 2015   175
G.R. Hubbell et al., *Remote Observatories for Amateur Astronomers*, The Patrick
Moore Practical Astronomy Series, DOI 10.1007/978-3-319-21906-6_11

Can Citizen Scientists Still Discover Asteroids
Using a Remote Telescope? (Rob Matson)

Objective and Goal of Project

With the early 1990s discovery of the 180-km-wide Chicxulub crater in the Gulf of
Mexico, it was generally accepted that the extinction of 75 % of the Earth's species
was triggered by an asteroid impact. This revelation no doubt generated a sense of
urgency in cataloging potentially hazardous asteroids. In 1995, NASA's Jet
Propulsion Laboratory entered into a cooperative agreement with the U.S. Air
Force to use its 1-m Ground-based Electro-Optical Deep Space Surveillance
(GEODSS) telescope at Haleakala, Maui, to search for near-Earth objects (NEO).
This campaign was called the Near-Earth Asteroid Tracking program (NEAT).
Beginning in 2000, NEAT in Maui transitioned to a nearby 1.2-m telescope and a
sister 1.2-m telescope was added in 2001 at Mount Palomar to search for NEOs. As
part of NEAT, a software tool named SkyMorph was developed to support targeted
searches within the growing NEAT image archive. This tool's primary purpose was
to help identify pre-discovery images of newly discovered asteroids and comets.
This is a very valuable function of the archive because it allows immediate update
on the initial orbits of newly discovered NEOs so that they will not become lost.

Perhaps an unanticipated consequence of NASA making SkyMorph and the
NEAT archive available online was that amateurs began finding new objects missed
by NEAT's automated search algorithms. I started mining the NEAT archives in
2003 and after a steep learning curve, began to find new asteroids, estimate their
initial orbits, and then use those orbit guesses to locate additional images on other
nights in the archive. Once I found and measured all possible nights for an object,
I forwarded my detailed observations (technically referred to as astrometry) to the
Minor Planet Center (MPC) in Cambridge, MA—the official clearinghouse for
asteroid discoveries. In some cases, the astrometry matched a known object, but
when it did not, I was issued a new preliminary designation.

I ultimately submitted about 1200 asteroids over a period of many years.
However, quite possibly because of the volume of new submissions by myself and
a half dozen other prolific NEAT sleuths, the MPC altered its rules for archive
finds. Instead of finders receiving discovery credit (and naming rights), the NEAT
program was credited from that point forward. With little motivation to spend additional hours padding NEAT's already impressive discovery tally (39,911 numbered
asteroids as of April 2015), I discontinued that work.

Still, this turned out to be a fortuitous change, because it allowed me to discover
the Sierra Stars Observatory Network (SSON). Instead of scouring another instrument's archives for missed objects, here was an opportunity to actually command a
telescope to find them! But was SSON up to the task? In this modern era of wide-
field-of-view survey instruments such as the Catalina Sky Survey (CSS), the
Panoramic Survey Telescope and Rapid Response System (Pan-STARRS), and even
an infrared, space-based asteroid-hunting satellite (NEOWISE), it seemed a daunting

task to discover a new minor planet with comparatively modest equipment. After all, any asteroid bright enough to be detectable with even a good-sized aperture surely would not be missed by one or more of the large surveys. Or would it?

The main objective was to develop a strategy to overcome SSON's competitive disadvantages in aperture, field-of-view, and sensitivity to discover a new asteroid. Assuming this strategy proved successful, the overall goal was to exercise these techniques over a period of years to discover numerous minor planets in a variety of orbital groups.

Strategy

Asteroids exhibit a slight boost in brightness when they reach the opposition region of the sky, i.e., close to 180 degrees from the Sun. This region is highest in the night sky at local midnight, and for mid-latitude observing locations is much higher in the winter than in the summer. The higher the elevation angle, the lower the atmospheric attenuation and the greater the sensitivity. If there were no other factors, the optimum time to search for asteroids would be at local midnight on the Winter Solstice. Unfortunately, this also happens to be about the time and location that the plane of the Milky Way intersects the plane of the Solar System (known as the ecliptic). The density of stars in the galactic plane is an order of magnitude greater than at the galactic pole, so it is definitely a region to avoid when searching for dim asteroids.

Ultimately, a balance must be struck between avoiding this galactic interference and maximizing elevation angle. This tradeoff favors observations from mid-August to mid-November and, to a lesser extent, mid-January to mid-March, when the elevation angle of the opposition region is still fairly high in the sky but the galactic plane is far from the ecliptic.

Because new discoveries of dim objects usually require excellent conditions, before deciding to command a particular telescope on a given night, it is a good idea to check its seeing forecast. The online Clear Sky Clock is an excellent tool for exploring conditions at the various SSON sites. The SSON website has convenient links, including this one for the Markleeville, CA, location: http://cleardark-sky.com/c/SrrStrsObCAkey.html. A typical chart appears in Fig. 11.1.

Obviously, there is little point in submitting jobs when the prognosis for operating is poor (rain, overcast, high winds), and you are very unlikely to discover new asteroids in the days surrounding full Moon because the limiting magnitude will be poor. Cloud cover, transparency, and seeing are all important for maximum asteroid detection sensitivity, so you're hoping that those little boxes from 10 p.m. to 2 a.m. local time are as dark blue as possible.

Assuming you've identified a night with promising dark sky conditions, the next step is to see what those big surveys (e.g., CSS and Pan-STARRS) have been up to lately. Because they've got you outgunned in both aperture and the field they can search per night, you don't want to waste time and money re-imaging fields that they've recently covered. MPC's link for big survey sky coverage can be found at: http://www.minorplanetcenter.org/iau/SkyCoverage.html.

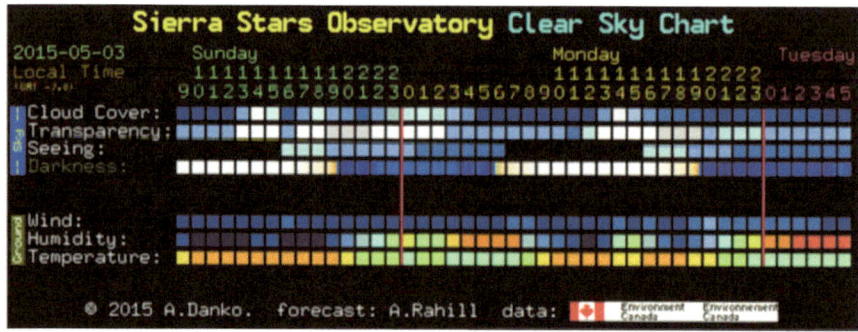

Fig. 11.1 Example Clear Sky Chart for SSO in Markleeville, CA

The default settings show you the most recent week of coverage for the entire celestial sphere where the instruments had a detection threshold of visual magnitude +18 or dimmer. You will want to modify these settings to show at least the last 10 days, opposition region only, and a magnitude limit of at least V = 19.5. Even at this dimmer limit, you may find that the surveys have been quite thorough by the time each new Moon rolls around. (The exception is during monsoon season, when many of the survey sites can get clouded out. For instance, CSS is located on Mount Lemmon in southern Arizona and thus is affected by the monsoon season, which runs from June 15 to September 30. Storms tend to peak between mid-August and mid-September—coincidentally the optimum time to be searching for new asteroids. What better time to search than when your greatest competition is offline?)

Figure 11.2 is an example chart showing the portions of the opposition region that were surveyed down to magnitude 19.5 during the 10-day period from April 27, 2015, to May 6, 2015.

The vertical dashed line near the center of the plot represents opposition. The gray curve looping from bottom right to upper left represents the galactic plane; the other gray curve passing through the group of red squares right of center represents the ecliptic plane. In deciding what part of the sky to image, you are looking for "exploitable holes" that are near the ecliptic, far from the galactic plane, and close to opposition. For instance, one promising region in this plot is between RA 13 h and 14 h, Declination 0 to +10 degrees.

With the original SSO telescope in northern California, you can routinely reach 20th magnitude (or even dimmer during particularly good seeing) with 2-min exposures. (This is the standard exposure duration I've used for most of my SSO asteroid searches.) To reliably detect asteroids, each field should be imaged four times at regular intervals. For Main Belt asteroids in the opposition region, images should be spaced at least 15–30 min apart to allow time for objects to move an appreciable distance between frames. The reason for four images is to hedge against an asteroid crossing a star in one of them, preventing accurate measurement. A minimum of three positions are needed to distinguish an asteroid from noise and allow the computation of a crude preliminary orbit.

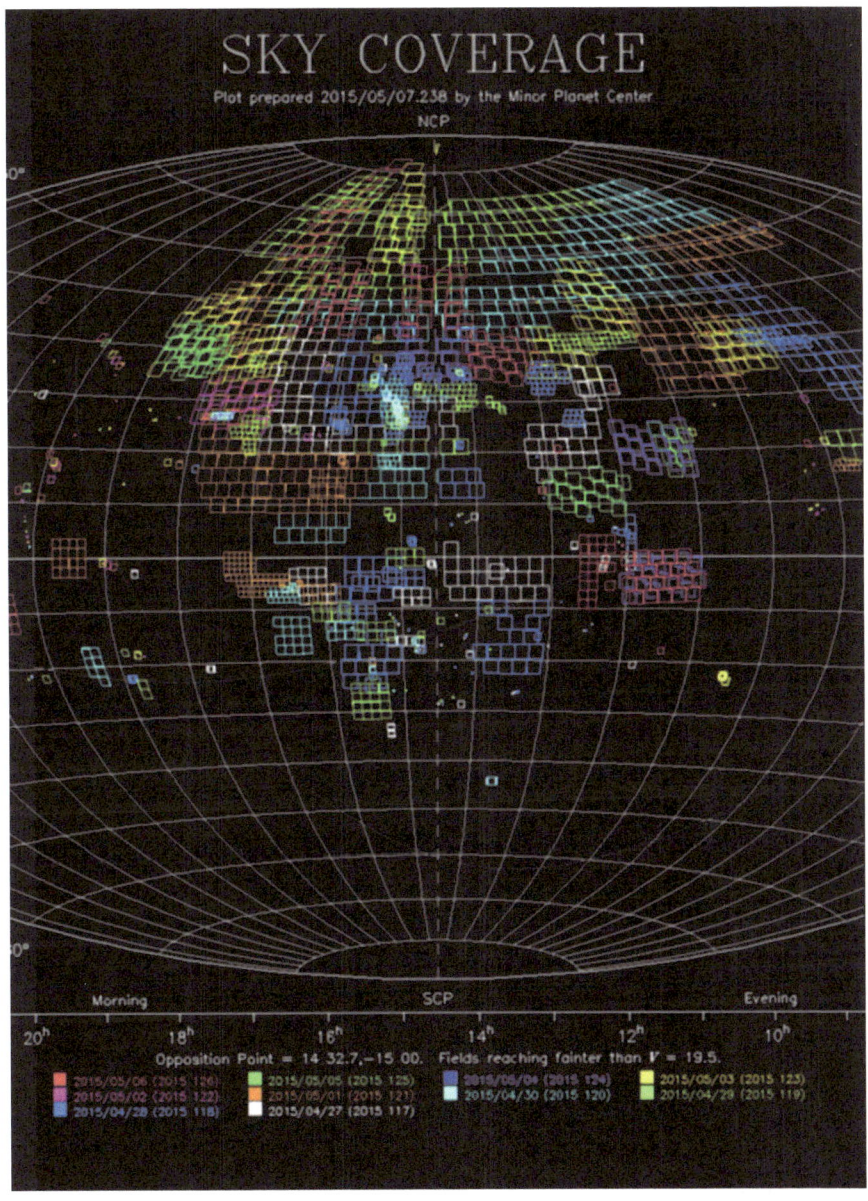

Fig. 11.2 Big survey sky coverage in the opposition region

Results

Anxious to attempt my first asteroid discovery with SSO, I ignored my own advice about waiting for the ecliptic to rise higher in the southern sky and started submitting jobs in mid-May 2009. After finding several surprisingly dim (but known) asteroids in the first few nights, I chose a small area in Libra that was devoid of known minor planets. Despite the non-ideal sky location, SSO's small 21-arc-minute-square field held not one but three new 20th magnitude asteroids (one of which I actually didn't spot until revisiting the images 3 months later). Not a week into the project and my asteroid discovery objective was already in sight.

However, finding a new asteroid in a single night's images is not enough for discovery credit: you must follow up with a second night, preferably within a few days. Knowing where a new asteroid will be on some future date requires an educated guess at the orbit and a method to propagate it.

Fortunately, you do not have to be a celestial mechanics expert to do this. Bill Gray's FindOrb orbital propagation software (http://www.projectpluto.com/find_orb.htm) is an excellent tool for creating a preliminary orbit and generating search ephemerides for follow-up observations. Main Belt asteroids move slowly enough from night to night that after 2 or 3 days, even if your initial orbit guess is very wrong, your quarry will likely be within 5 arc-minutes of your prediction—well within the telescope's field of view. Figure 11.3 shows my first SSO asteroid discovery, 2009 KW, in a three-frame stack of the second night recovery images. (The bright streak at the bottom of the image is a manmade satellite that passed through the field of view during one of the three frames—a not-uncommon occurrence.)

After 6 years' experience with SSON, I can say the project has been a resounding success. Table 11.1 lists 57 asteroids that were discovered, their discovery dates, the orbital group that each asteroid belongs to (when known), and a color-coded status (green for SSON-credited discoveries, red for SSON finds that ultimately got linked and credited to an earlier observer).

Summary of Discovery Steps

Preparation:

1. Check Clear Sky Clock for clouds/seeing at the various SSON telescope sites.
2. Determine the approximate right ascension and declination of the opposition region for the intended night, and the location of the ecliptic. Focus on locations on the celestial sphere that are within a few hours right ascension of the opposition point, and select approximate declination based on a tradeoff among proximity to the ecliptic, avoidance of the galactic plane, and favoring higher elevations over lower ones.

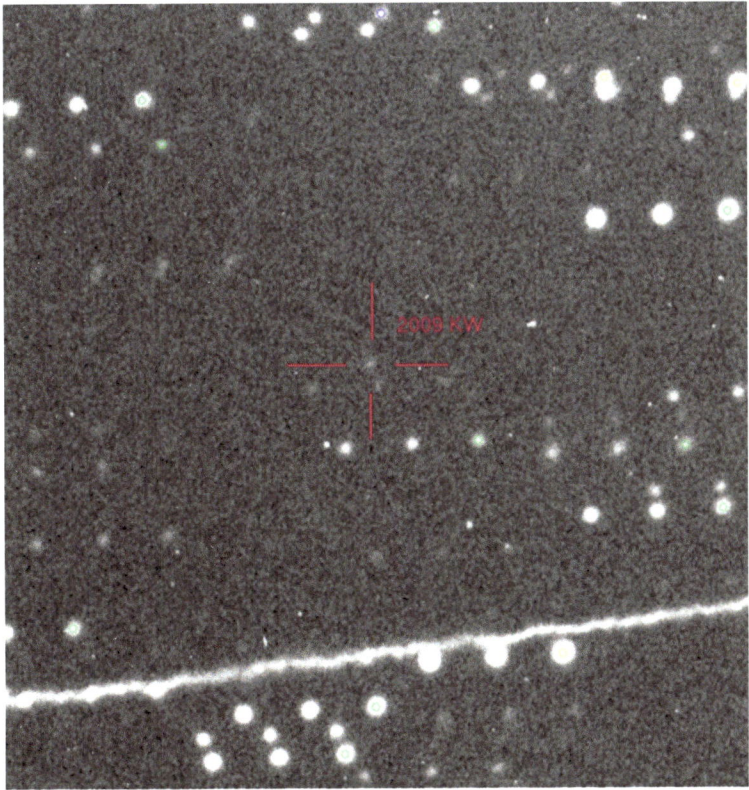

Fig. 11.3 R. Matson SSO asteroid discovery #1, 2009 KW, in second night's frame stack

Table 11.1 Asteroids discovered (or rediscovered) by Matson using SSON

Discovery number	Initial designation	Discovery date	Group	Current name/status
1	2009 KW	5/16/2009	Main Belt I	(296503) 2009 KW
2	2009 KL3	5/23/2009	Flora	(296506) 2009 KL3
3	2009 KM3	5/23/2009	Main Belt IIb	(312594) 2009 KM3
4	2009 OC2	7/20/2009	Main Belt I	(325423) 2009 OC2
5	2009 OF4	7/24/2009	Main Belt I	(379147) 2009 OF4
6	2009 OB15	7/28/2009	Main Belt IIb	2009 OB15
7	2009 OC15	7/28/2009	Main Belt IIIb	(264006) 2009 OC15
8	2009 OD15	7/28/2009	Main Belt IIIb	2009 OD15
9	2009 KT30	5/16/2009	Main Belt I	(316092) 2009 KT30
10	2009 QF7	8/16/2009	Main Belt I	2009 QF7
11	2009 QG7	8/19/2009	Main Belt I	2009 QG7
12	2009 QA9	8/19/2009	Main Belt II	2009 QA9

(continued)

Table 11.1 (continued)

Discovery number	Initial designation	Discovery date	Group	Current name/status
13	2009 QB9	8/19/2009	Flora	2006 WK121
14	2009 QC9	8/20/2009	Mars-crosser	2009 QC9
15	2009 QD9	8/20/2009	Main Belt IIIb	2009 QD9
16	2009 QL28	8/22/2009	Main Belt IIIa	2009 QL28
17	2009 QZ31	8/20/2009	Main Belt I	2009 QZ31
18	2009 QS32	8/25/2009	Flora	2009 QS32
19	2009 QT32	8/25/2009	Main Belt IIb	2009 QT32
20	2009 QM34	8/27/2009	Main Belt I	(372393) 2009 QM34
21	2009 QD36	8/27/2009	Main Belt II	2009 QD36
22	2009 QG36	8/27/2009	Mars-approacher	2009 QG36
23	2009 QJ36	8/28/2009	Main Belt II	(423624) 2005 WZ156
24	2009 SP1	9/17/2009	Maria	2009 SP1
25	2009 SC20	9/22/2009	Main Belt IIIb	2009 SC20=2014 SR302
26	2009 SD20	9/22/2009	Main Belt IIIb	2004 TN77
27	2009 SG20	9/22/2009	Main Belt IIIb	2009 SG20
28	2009 SQ98	9/23/2009	Main Belt IIIb	(404959) 1998 RS13
29	2009 SZ167	9/25/2009	Main Belt IIIb	(321506) 2009 SZ167
30	2009 SA168	9/25/2009	Main Belt IIIb	2009 SA168
31	2009 UE129	10/29/2009	Trojan	2009 UE129
32	2009 UF129	10/29/2009	Main Belt IIb	2009 UF129
33	2009 VM	11/7/2009	Main Belt II	(332772) 2009 VM
34	2009 WC104	11/23/2009	Main Belt IIb	(423243) 2004 TK88
35	2009 WD104	11/23/2009	Main Belt IIb	2009 WD104
36	2010 ES31	3/11/2010	Main Belt IIIb	2010 ES31
37	2010 RT26	9/3/2010	?	2010 RT26
37	2010 VK183	11/12/2010	Flora	2010 VK183
39	2011 OT15	7/27/2011	Main Belt II	2011 OT15
40	2011 OS17	7/29/2011	Main Belt I	2011 OS17
41	2011 OT17	7/29/2011	Main Belt I	2004 TE244
42	2011 OU17	7/29/2011	Main Belt IIIb	2011 OU17
43	2011 QB22	8/25/2011	Main Belt IIb	(350870) 2002 OF37
44	2011 QC22	8/25/2011	Main Belt I	2011 QC22
45	2011 QD22	8/25/2011	Flora	2011 QD22
46	2011 QE22	8/25/2011	Main Belt IIIb	2011 QE22
47	2011 QL51	8/31/2011	Main Belt I	2 011 QL51
48	2011 RA1	9/3/2011	Main Belt IIb	2011 RA1
49	2011 UZ290	10/30/2011	Main Belt I	2009 BW131
50	2013 EZ35	3/7/2013	Main Belt IIIb	2013 EZ35
51	2013 EA36	3/7/2013	Mars-crosser	2013 EA36
52	2013 EA41	3/11/2013	Trojan	2013 EA41
53	2013 EB41	3/11/2013	Main Belt IIIb	2013 EB41
54	2013 EC41	3/11/2013	Hungaria	2013 EC41
55	2014 YX35	12/23/2014	?	2014 YX35
56	2015 CQ3	2/4/2015	Main Belt II	2015 CQ3
57	2015 DV92	2/16/2015	?	2015 DV92

3. Use MPC's Sky Coverage Plot to check which fields have been imaged within the last 7–10 days by the major survey instruments, and favor locations that have either not been covered or were only imaged to modest limiting magnitude (e.g., Mv +19).
4. Submit SSON jobs for one or more sky locations, repeating observations at each location four times with a cadence of 15, 20, or 30 min depending on the expected angular rate of the desired target(s). Choose exposure times from 60–180 s depending on the telescope, expected seeing, and cost-balance between total fields imaged and the limiting magnitude of each field. Always choose the "clear" or "no filter" option for the filter band.

Processing:

1. Wait for email announcement from SSON that your images are ready for download (usually by noon the following day).
2. Use FileZilla ftp software to access and download the ZIP files of your images.
3. UnZIP the images, organizing them into appropriate folders.
4. Start Herbert Raab's Astrometrica (or similar astrometric software for displaying, blinking, and accurately measuring CCD images), and load the first group of images.
5. Astrometrically align (register) the images, and begin blinking them.
6. Systematically scan the blinking images looking for moving objects.
7. For each moving object found, assign a six-character preliminary designation (e.g., MAT001), and record the positions and magnitudes.

Analysis and Follow-Up:

1. Go to the MPC's MPChecker link (http://scully.cfa.harvard.edu/cgi-bin/checkmp.cgi) and check all objects found for matches in MPC's database.
2. For matches to known objects, change the preliminary designation to that of the known object and submit the new observations to MPC.
3. For first-time objects with no MPC match, run FindOrb to compute a preliminary (typically Väisälä-type) orbit. Check the weather/seeing conditions at the various SSON sites for promising opportunities for follow-up observations. Depending on the angular speed of the new object, it may be necessary to reacquire it in the next day or two to avoid losing it. If it is moving very fast and is thus a suspected NEO, email the single-night observations to the MPC immediately. Otherwise, try to submit a new job to SSON in the next 2–6 days for follow-up observations.
4. Repeat processing steps for second-night images. For objects successfully recovered, use FindOrb to confirm that your assumed linkage is correct and that the residuals seem reasonable.
5. Submit the two-night linked observations to the MPC and cross your fingers that it's a new object!

Benefit Derived from Using Remote Observatory Facilities

This asteroid discovery project demonstrated that with strategic planning and perseverance, amateurs can use remotely commanded telescopes to not only successfully discover new minor planets, but do so at modest cost.

Facility Equipment Used

Three of the telescopes in SSON's five-telescope network were used for this project:

- Rigel (Sonoita, AZ): 0.37-m, F/14 classical Cassegrain telescope, Kodak® KAF-16803 4096 by 4096 CCD, binned 2 by 2, clear filter.
- Sierra Stars Observatory (Markleeville, CA): 0.61-m, F/10 classical Cassegrain telescope, Kodak® KAF-09000 3056 by 3056 CCD, binned 2 by 2, clear filter.
- Mount Lemmon SkyCenter Schulman Telescope (Mt. Lemmon, AZ): 0.81-m, F/7 Ritchey-Chretien telescope, Kodak® KAF-16803 4096 by 4096 CCD, binned 2 by 2, luminance filter.

Epilogue

In December 2014, I decided to try using SSON for a new project: deep-sky, color astrophotography. For my first attempt, I chose the photogenic Horsehead Nebula in Orion, which at the time was optimally located for northern hemisphere observers. Imagine my surprise when I opened my quintet of luminance images the next day and spotted a dim object creeping toward the Horsehead's mane! Follow-up images a few days later confirmed it was unknown, and MPC assigned 2014 YX35 to my 55th asteroid discovery with SSO (Fig. 11.4).

About Rob Matson

Rob Matson is a space scientist and amateur astronomer. He is a successful hunter of asteroids, comets, and meteorites. Minor planet 73491 (Robmatson) was named after him. Rob lives with his wife Lisa in Newport Coast, CA.

Fig. 11.4 Five-image stack-and-track of 2014 YX35 approaching Horsehead Nebula on December 23, 2014

Objective and Goal of Project

As of May 2015, more than 12,600 near-Earth asteroids (NEA) have been identified. Unfortunately, physical characterization has been lagging the discovery effort and only about 15 % of NEAs have any physical parameters measured for them. Some types of observations, such as rotational light-curve and phase function photometry, require large blocks of time, often across multiple nights. The required cadence for these observations is often difficult to obtain on large professionally operated telescopes. To address the need for physical characterization and to increase access to more telescopes, the OSIRIS-REx Target Asteroids! program was established as a citizen science endeavor to characterize NEAs.

The program leverages the still relatively untapped worldwide resources of modest-sized telescopes, including many remotely operated facilities, to contribute to our understanding of NEAs.

The Target Asteroids! citizen science program is a part of the Communications and Public Engagement activities of the NASA OSIRIS-REx asteroid sample return mission, which will launch in 2016, collect samples of potentially hazardous NEA (101955) Bennu in 2019, and return samples to Earth in 2023. The main goals of Target Asteroids! are twofold:

- Increase our knowledge of the NEA population with special emphasis on potential space exploration mission targets, whether human or robotic.
- Encourage amateur astronomers to develop the expertise and tools to contribute astrometric, photometric, and spectroscopic observations of NEAs.

The science objectives of Target Asteroids! focus on the study of three populations of asteroids:

- **Spacecraft accessible asteroids.** The low delta-V orbits of such objects are the most conducive to spacecraft rendezvous and the return of material to Earth. The number of such objects for exploration is low, and precious few have been sufficiently characterized, limiting our ability to plan missions.
- **Analogs to carbonaceous asteroids such as Bennu.** Missions such as OSIRIS-REx and Hayabusa-2, a Japanese-led mission to asteroid (162173) 1999 JU3, focus on studying carbonaceous asteroids. Such asteroids are believed to be rich in carbon, organics, and volatiles like water, in short, the material of life. Although Bennu and 1999 JU3 are very well characterized, there are still gaps in our understanding. This is especially true for photometry at very low (less than 15 degrees) and very high (greater than 100 degrees) phase angles (the Sun-asteroid-observer angle).
- **Inner Main Belt asteroids that may be related to Bennu and 1999 JU3.** The observed Main Belt asteroids are members of asteroid families identified by Walsh et al. (2013) as possible sources of carbonaceous low delta-V NEAs. Not only do the Main Belt asteroids teach us more about the origins of NEAs and their dynamical pathways into the inner Solar System, but they also provide brighter targets to observe with small telescopes. Such observers can use techniques learned studying brighter objects to examine fainter NEAs when they acquire access to larger telescopes.

Results

While observers with smaller telescopes cannot produce high signal-to-noise ratio (SNR) observations for faint objects, they do have the advantage of permitting many observations. Even lower SNR observations are useful if made often over different observing geometries. This is why our main goal is to acquire photometry of asteroids over a large range of phase angles. The relationship between the brightness of an asteroid and its phase angle is called its phase function. By obtaining numerous low-precision photometric observations over a range of solar phase angles, direct measurements of the phase function, absolute magnitude, color, and rotation period, as well as indirect measurements of the taxonomy, albedo, and size, can be determined for asteroids with modest-sized telescopes.

Determining how the light-scattering properties of the surface of an asteroid change with different observing geometries provides us with many important parameters. The slope of the phase function is directly correlated with albedo (Belskaya and Shevchenko 2000; Oszkiewicz et al. 2012; Hergenrother et al. 2013). Highly reflective asteroids with albedos of about 0.4 or greater have phase slopes of about 0.02 magnitudes per degree of phase angle, while dark asteroids (such as Bennu) with low albedos (on the order of about 0.05 or less) have phase slopes of about 0.04 magnitude per degree of phase angle (Fig. 11.5).

Modeling the shape of the phase function and extrapolating it to a phase function of 0 degrees produces a value for the absolute magnitude (H). (When referring to absolute magnitude of a Solar System object, we mean the brightness it would have if observed from the Sun and the object were 1 AU away.) Combining our knowledge of H and albedo yields an estimate of the size of the asteroid. If observations are acquired in different broadband filters (such as BVRI) at varying phase angles, changes in the color of the asteroid with phase angle can be derived. Colors can be used to determine the taxonomy and detect any taxonomic changes at different observing geometries.

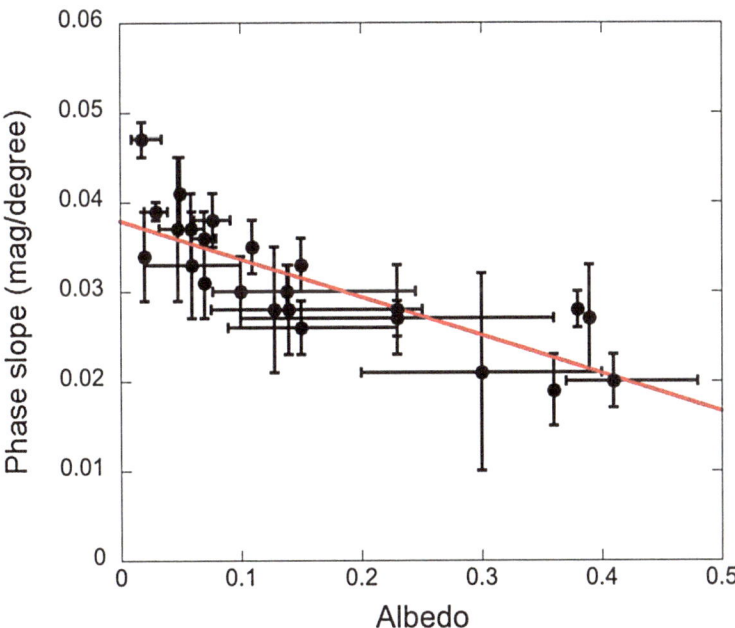

Fig. 11.5 Correlation between the slope of the phase function and albedo for NEAs with albedo measurements (Data are from Hergenrother et al. (2013))

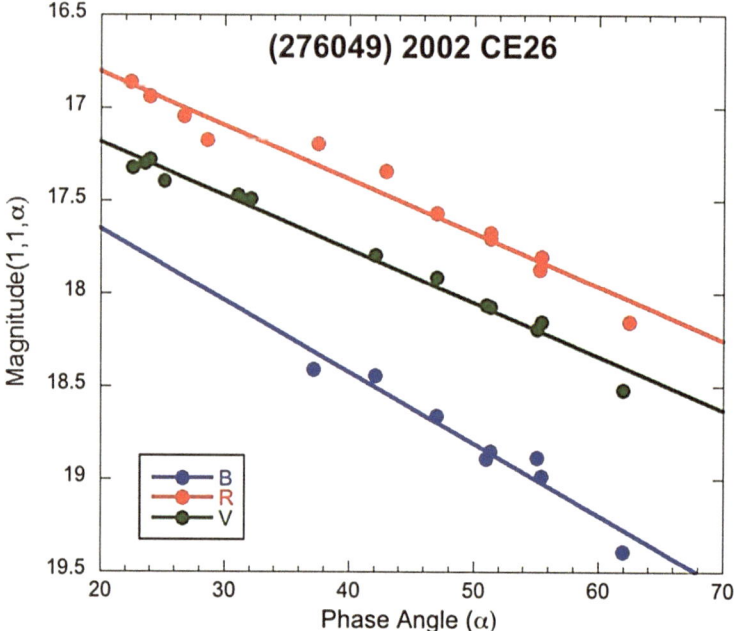

Fig. 11.6 The binary (possibly triple) system NEA (276049) 2002 CE26 was extensively observed by many Target Asteroids! observers during summer 2014. The figure shows B, V, and R band phase functions and possible color changes especially in the B-V index. Each point is the mean of the light-curve photometry obtained by a single observer on a single date

Over the past 2 years, the Target Asteroids! program has produced data for more than 50 NEAs and Main Belt carbonaceous asteroids. Target Asteroids! asks its observers to collect data for dedicated observing campaigns. Data are then uploaded to a server at the University of Arizona and reduced. The astrometry/photometry software package Astrometrica and MPO Canopus© are used to reduce the BVR photometry (Raab 2012; Warner et al. 2011). The software compares the brightness of the asteroids with photometric reference stars within the field. A circular aperture of the same size is used for both the target object and the reference stars in each field. (USNO CCD Astrograph Catalog 4 (UCAC4) stars with solar-type colors are used as photometric references (Zacharias et al. 2013)). Some of our more experienced observers have used Astrometrica and MPO Canopus© to reduce data on their own. In these cases, both their images and reduced photometry are archived.

Observations of NEA (276049) 2002 CE26 highlight the quality of data produced by Target Asteroids! members equipped with small (less than 0.6-m) telescopes. During summer and fall 2014, 2002 CE26 peaked at an R magnitude of about 14. Figure 11.6 shows a phase curve for 2002 CE26 consisting of B, V, and R band data taken between V magnitudes of 13.6 and 16. All the B band and some

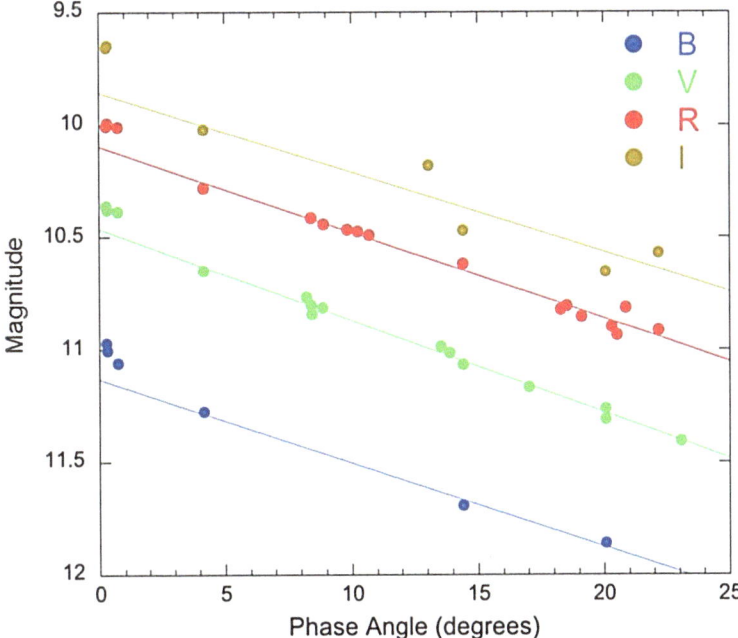

Fig. 11.7 Preliminary analysis of B, V, R, and I photometry of Main Belt asteroid (142) Polana shows color changes with respect to phase angle

of the V and R band data were acquired with remotely operated telescopes. The difference in the slopes of the phase functions in each color is a sign of color changes at different phase angles.

Color photometry of the inner Main Belt carbonaceous asteroid (142) Polana shows the consistency between multiple observers. The B, V, and R phase curves in Fig. 11.7 were produced by nine separate observers. (Two used remotely operated facilities.) Some of the data were taken in V and R filters while other data were unfiltered and transformed to the V or R colors. The tight correlation shows that data from many different observers can be merged into a high-quality result that can be used to detect subtle changes in the spectrum of asteroids observed at different observing geometries (Fig. 11.8).

Benefit Derived from Using Remote Observatory Facilities

When formulating the program, we had to ask the question: how can small telescope users expand our understanding of asteroids in an era of very large aperture telescopes? Some measurements, such as taxonomic type, color, and albedo, can be

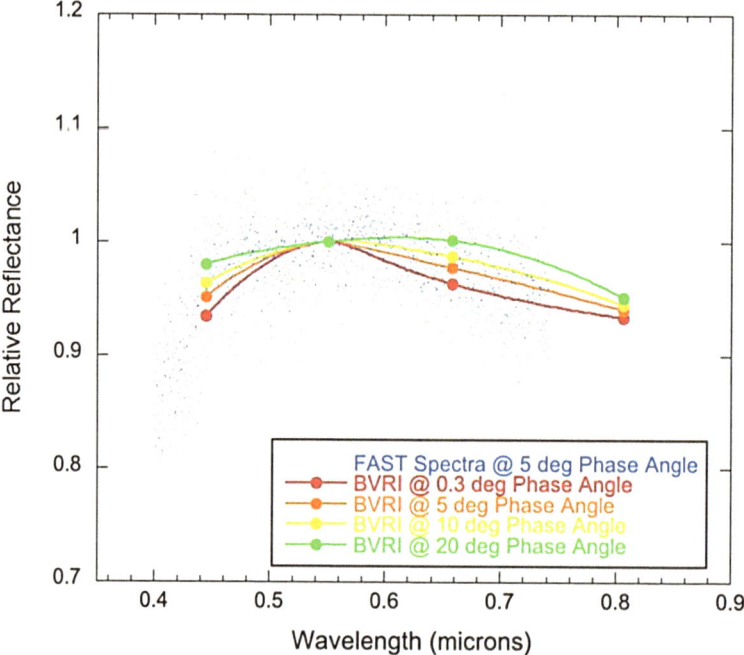

Fig. 11.8 The spectra of Polana are observed to flatten with increasing phase angle (*bluer* at wavelengths shorter than 0.55 μm and *redder* at wavelengths longer than 0.55 μm). All of the observed colors are consistent with carbonaceous taxonomies, although the exact taxonomy changed with increasing phase angle

determined in mere minutes with large telescopes. Other observations, such as rotation period determination, can take many hours or days, which may be difficult to obtain at large facilities. To measure an object's phase function and to detect changes in color or other properties over a large range of phase angles, observations taken across many nights are required. The time to make these measurements can span from weeks to months.

Target Asteroids! is as much about encouraging citizen scientists to observe asteroids as it is about the study of the asteroids themselves. Remote facilities represent an option for amateurs to be involved even if they don't have a telescope or dark skies. They also allow observers to observe with telescopes beyond their capacity to own or operate, provide access to science-grade instrumentation and enable observations of objects otherwise invisible from their home locations.

For science, remote facilities are very important as gap-fillers. Most Target Asteroids! data are taken from the backyards of our observers. In an ideal world, our observers would be spread evenly across the globe covering all longitudes and latitudes. The reality is that the vast majority of them are located in the northern hemisphere and mostly in North America and Europe. Access to remote telescopes at other longitudes and latitudes allows observation of more objects and, in some cases, even comes close to allowing continuous coverage. An additional benefit of many remote telescopes is the availability of photometric filters (BVRI and Sloan filters) that most backyard observers lack.

Facility Equipment Used

Target Asteroids! has received imaging and spectroscopic data taken with telescopes as small as a 0.13-m reflector to the Vatican VATT 1.8-m telescope located on Mount Graham in southeast Arizona. The majority of the submissions have come from 0.3-m to 0.6-m telescopes. Remotely operated telescopes used for the program include the Sierra Stars 0.6-m (equipped with FLI ProLine PL09000 and located in California), 0.5-m (equipped with a SBIG STL6303E CCD and located at Warrumbungle, Australia) and 0.3-m (equipped with a FLI ProLine PL16803 CCD and located in Arizona), and the iTelescopes.net 0.7-m and 0.5-m (both equipped with FLI ProLine PL16803 CCD and located at Siding Spring, Australia). Most of the individual telescopes were selected not only based on aperture size and location but also because of their use of photometric B, V, R and I filters.

About Carl Hergenrother

Carl Hergenrother is a staff scientist at the University of Arizona's Lunar and Planetary Laboratory. His research interests focus on the telescopic study of asteroids and comets. In particular, he has analyzed the rotation and spectral properties of very small near-Earth asteroids and comet-asteroid transition objects. As an OSIRIS-REx science team member, Carl is tasked with making observations of the target asteroid Bennu that can be directly compared with remote observations made from Earth, planning spacecraft operations at the asteroid, and leading an amateur-professional program to study Bennu-like asteroids with small telescopes.

Before working on the OSIRIS-REx Mission, Carl was a member of the Catalina Sky Survey (CSS). During his tenure with CSS, he discovered four comets and numerous near-Earth asteroids. He has worked for the Minor Planet Center determining the orbits of newly discovered asteroids.

Carl is an avid backyard astronomer (both visually and with CCD imagers). He enjoys observing meteor showers and comets. He is the Acting Coordinator of the Comet Section of the Association of Lunar and Planetary Observers, an associate editor for the *International Comet Quarterly*, a handling editor for the International Meteor Organization, and Secretary of the American Meteor Society.

References

Belskaya IN, Shevchenko VG (2000) Opposition effect of asteroids, *Icarus* 147, p. 94–105

Hergenrother CW, Hill D (2013) The OSIRIS-REx Target Asteroids! project: A small telescope initiative to characterize potential spacecraft mission target asteroids, *Minor Planet Bulletin* 40, p. 164–166

Hergenrother CW, Nolan MC, Binzel RP, Cloutis EA, Barucci MA, Michel P, Scheeres D, d'Aubigny CD, Lazzaro D, Pinilla-Alonso N, Campins H, Licandro J, Clark BE, Rizk B, Beshore EC, Lauretta LS (2013) Light curve, color and phase function photometry of the OSIRIS-REx target asteroid (101955) Bennu, *Icarus* 226, 663–670.

Oszkiewicz DA, Bowell E, Wasserman LH, Muinonen K, Penttilä A, Pieniluoma T, Trilling DE, Thomas CA (2012) Asteroid taxonomic signatures from photometric phase curves, *Icarus* 219, 283–296.

Raab H (2012). Astrometrica: astrometric data reduction of CCD images. Astrophysics Source Code Library, record ascl:1203.012.

Walsh K, Delbo M, Bottke W, Vokrouhlicky D, Lauretta D (2013) Introducing the Eulalia and New Polana asteroid families: re-assessing primitive asteroid families in the inner Main Belt, *Icarus* 225, 283–297.

Warner BD, Stephens RD, Harris AW (2011) Save the light curves. *Minor Planet Bulletin* 38, 172–174.

Zacharias N, Finch CT, Girard TM, Henden A, Bartlett JL, Monet DG, Zacharias MI (2013) The fourth US Naval Observatory CCD astrograph catalog (UCAC4). AJ 145, id. 44.

Chapter 12

Remote Astronomy Education Projects

College and university professors are starting to use remote observatory facilities as part of astronomy laboratory work projects. Students acquire images with large professional telescopes in remote dark-sky sites that are impossible to obtain from light polluted campuses. The growth of remote observatories, both proprietary and commercial, will enable more students to do these types of "hands-on" astronomy imaging projects.

This chapter describes three different projects in which professional astronomers use remote observatories for education and research programs that enable thousands of students around the world to participate.

Remote Astronomy at Mesa Community College (Kevin Healy)

Objective and Goal of Project

While many college students and members of the public appreciate the beauty of an astronomical image, few understand the concepts behind imaging. The details of astronomical imaging software can be complex, but basic observing concepts such as rise and set times and basic analysis concepts such as the meaning of pixel values, image alignment, and color balance can be mastered after a few hours of practice.

© Springer International Publishing Switzerland 2015   193
G.R. Hubbell et al., *Remote Observatories for Amateur Astronomers*, The Patrick
Moore Practical Astronomy Series, DOI 10.1007/978-3-319-21906-6_12

We have developed a pair of telescopic observing laboratories at Mesa Community College that focus on astronomical imaging. The first laboratory exercise introduces concepts related to astronomical images and observation planning, while the second laboratory exercise asks the students to process an image set of their choosing to produce a final result and report on their findings to fellow students. Our students use the Mira® AL software package to view, edit, and process their images. Mira® AL is a simplified version of the Mira® software specifically designed for student users.

In the first laboratory, students learn to load FITS image files and inspect image headers for details such as observing date and time, filter setting, and exposure time. They also learn to recognize image artifacts such as tracking errors and cosmic rays. A short tutorial introduces the process of image alignment and then students align an example image set and inspect it for an asteroid moving in the starry sky. Students also examine an image set recorded through color filters and construct a three-color-composite image.

The first laboratory ends with students planning two telescopic observations with the use of planetarium software. The first target to be chosen is an asteroid or Kuiper Belt object. The second target is a nebula or galaxy from the Messier or NGC catalogs. In planning their observation, students must consider which objects are up at night during the semester, and they must choose objects that have relatively small apparent diameters (to fit in the camera's field-of-view) and relatively bright apparent magnitudes (to make the exposure time manageable).

Between the first and second laboratories, instructors pool all student object requests and create an observing list from the most popular Solar System and deep sky objects. Observing requests are made using the telescopes of the Sierra Stars Observatory Network (SSON). Solar System objects are typically scheduled for six observations at 1-h intervals. Main Belt asteroids can be detected visually in images as short as 30 s while the larger Kuiper Belt objects can be detected visually in images of 60-s exposure time. Nebulae and galaxies are observed for 200–300 s per filter, with red-green-blue or red-visual-blue sequences chosen based on telescope availability. Based on the emission characteristics of the object, a hydrogen-alpha image may also be captured. These exposure times are long enough to record a recognizable image of the object without requiring stacking and co-adding multiple images that would add complexity to the students' image analysis.

With the image data archived, the students are ready for the second imaging laboratory exercise. In this exercise, student groups choose either a small Solar System object or a deep sky object from the available image sets that semester. The students then inspect their image sets for image artifacts such as star trails caused by telescope tracking errors or out-of-focus exposures. Instructors monitor the images as they are collected and reschedule missed observations, but this step allows students to see the kinds of things that can go wrong and identify the problem.

Students who choose a Solar System object for the second laboratory exercise place the images in chronological order and record image properties such as date and time of observation, filter setting, and exposure time. The images are then aligned to a common set of reference stars, and the images are played in sequence to reveal the motion of the asteroid or Kuiper Belt object against the background stars.

Sometimes students will inadvertently choose the moving object as one of the alignment reference stars. Instructors can use this as a teaching opportunity to explain the alignment method. Most student groups immediately recognize what has happened.

Students who choose a deep sky object for the second laboratory exercise record image properties such as date and time of observation, filter setting, and exposure time. The images are aligned to a common set of reference stars and placed in RGB or RVB order. Then, median pixel values are measured for portions of the sky and the object, and those values are entered into the Merge RGB tool. If students are satisfied with the color balance of the resulting image, they are done. If not, the minimum (sky) and maximum (object) pixel values are adjusted until a satisfactory result is achieved.

After the image analysis process is complete, students investigate the target of their observations. Details such as distance, size, type of object, year of discovery, and interesting facts are gathered. When all student groups are ready, each group presents its final animation (for Solar System objects) or color image (for deep sky objects) to the class. Each student is responsible for providing his or her share of the group report.

Results

Typical imaging results are presented in Figs. 12.1 and 12.2. Figure 12.1 shows the motion of dwarf planet 136472 Makemake during a 5-h interval. Typically, students view the motion of the small Solar System body as an animation in Mira® AL. Here the images have been median-combined to show the changing position of Makemake during the interval of exposures as a short trail (shown in the red box) during the 5-h interval. Images were 60 s in length through the clear filter and were separated by 1 h. North is up and east is to the left. The field of view of the image is 21 arc-minutes square.

Figure 12.2 shows a typical three-color-composite image produced by students. This image shows M83, a spiral galaxy in the constellation Hydra. No processing was done to the images other than aligning them to common reference stars. (No processing to remove cosmic ray artifacts or bad columns.) Even so, the color image shows a wealth of features such as the dust lanes and blue color of the spiral arms that contrast with the yellow color of the galaxy's central bulge. Images were 300 s in length through R, V, and B filters. While the exposure times are short compared with typical high-quality astrophotography, this color image reveals major structures in the galaxy: dust lanes, blue color in the spiral arms, and yellow color in the central bulge. Note two distant galaxies faintly visible to the left of M83. Red, green, and blue speckles are artifacts caused by cosmic ray hits to the CCD. North is up and east is to the left. The field of view of the image is 21 arc-minutes square.

The two telescope imaging laboratory exercises described above have become our most popular exercises. Students make decisions on collecting real data and have the opportunity to examine and analyze the images they have requested.

Fig. 12.1 Six hours of orbital motion of dwarf planet 136472 Makemake captured by the SSON 24-in. telescope on the night of March 27, 2015

Benefit Derived from Using Remote Observatory Facilities

When this laboratory concept was developed, our initial goal was to provide students with the opportunity to complete observing projects all the way from telescope and CCD operation to image processing and analysis. However, our experiences with balky CCD cameras, a shaky rooftop observing deck, and poor imaging conditions inside the suburban light pollution dome of metropolitan Phoenix led us to look for alternatives.

Over the past 3 years, we have used the SSON telescopes to gather high-quality images for a nominal fee. We collect a $10 laboratory fee from each astronomy student and use a small fraction of this funding for each semester's imaging costs. With the addition of southern hemisphere telescopes, we are excited to think about projects involving the southern Milky Way and the Magellanic Clouds. We expect to continue our very successful partnership with SSON.

Fig. 12.2 A three-color-composite image of spiral galaxy Messier 83, captured by the SSON 24-in. telescope on the night of March 27, 2015

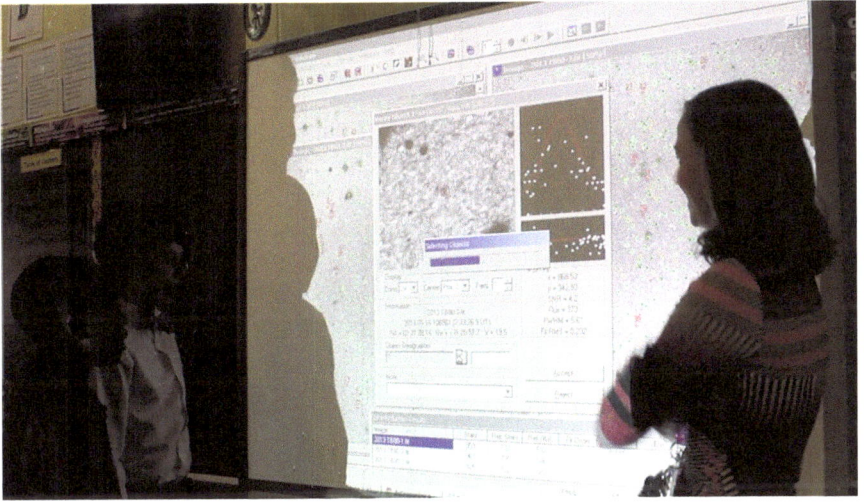

Fig. 12.3 A teacher and students learning about measuring star and asteroid positions using Astrometrica

Facility Equipment Used

We used the SSON 24-in. (0.6-m) in California to acquire most of our images for projects. In addition, we also used the SSON Rigel 14.5-in. (0.35-m) telescope and the SSON Mount Lemmon SkyCenter 32-in. (0.8-m) telescope to acquire images when the California telescope was unavailable because of extended bad weather.

About Kevin Healy

Dr. Kevin Healy is astronomy faculty and planetarium director at Mesa Community College in Mesa, AZ. He has a lifelong passion for astronomy education and public outreach. Kevin earned a Ph.D. in physics from Arizona State University, observing star-forming regions at visible, infrared, and radio wavelengths. Before graduate school, he earned a B.S. in physics at the New Mexico Institute of Mining and Technology and then worked as a data analyst at the National Radio Astronomy Observatory (NRAO) Array Operations Center, both in Socorro, NM. There Kevin supported visiting and staff astronomers who used the Very Large Array and the Very Long Baseline Array.

International Astronomical Search Collaboration (IASC) (Patrick Miller)

Objective and Goal of the Project

The International Astronomical Search Collaboration (IASC or "Isaac") is an online outreach program for high schools and colleges. Students make original discoveries of Main Belt asteroids (MBA), near-Earth objects (NEO), and trans-Neptunian objects (TNO).

Each year, 10,000 students from 500 schools located in more than 65 countries participate. On a daily basis, they receive image sets from professional observatories, provided to them online at http://iasc.hsutx.edu. The students use the software Astrometrica to search for, measure, and discover moving targets within the Solar System.

The measurements and discoveries are officially report to the Minor Planet Center (MPC, Harvard). The MPC is recognized by the International Astronomical Union (IAU, Paris) as the world's official body that maintains the database of all known minor bodies in the Solar System. The bodies include MBAs, NEOs, comets, and TNOs.

The professional observatories include:

• Pan-STARRS (Institute for Astronomy, University of Hawaii)
• Astronomical Research Institute (Westfield, IL)

- Sierra Stars Observatory Network (Markleville, CA)
- Catalina Sky Survey (Lunar & Planetary Laboratory, University of Arizona)
- Las Cumbras Observatory (Sutherland site, South African Astronomical Observatory)

In addition to giving high school and college students the unique opportunity to make original astronomical discoveries, IASC provides teachers in their classrooms with an online, hands-on educational tool. This tool has been integrated into science curriculum all over the world, with 85 % of the teachers returning each year to introduce new students to the program.

Astronomy educational outreach organizations worldwide use IASC to provide this unique opportunity to area schools and teams. They recruit schools and teams, and train the students and teachers. The following organizations have adopted IASC as one of their educational outreach services:

- Haus der Astronomie (Max Planck Institute for Astronomy; Heidelberg, Germany)
- Space Popularization Association of Communicators & Educators (New Delhi, India)
- Nojum Magazine (Tehran, Iran)
- Núcleo Interactivo de Astronomia (Lisbon, Portugal)
- National Astronomical Observatories of China (Beijing, China)
- Asociación Larense de Astronomía (Barquisimeto, Venezuela)
- Secondary School Council of Uruguay
- Luis Cruls Astronomy Club (Campos dos Goytacazes, Brazil)
- Astronomical Society of South Africa
- Astronomers Without Borders (Santa Barbara, CA)
- Space Generation Advisory Council (Vienna, Austria)

Results

Since IASC began in October 2006, there have been 1015 Main Belt asteroid discoveries. These discoveries have reached provisional status at the MPC. Of these, 33 have been numbered, cataloged, and named by their student discoverers. (From the time of the confirmation of an original discovery, it takes 6–10 years (or longer) for the asteroid to be numbered and cataloged. This delay in time is required for sufficient measurements to be made to fully determine the orbit.)

In one instance, students from Madisonville Junior High School (Madisonville, TX) discovered the Main Belt asteroid 2008 SE209. Their teacher, Denise Rothrock, led the students as they participated in the IASC search program. Meanwhile, Ms. Rothrock transferred to Madisonville High School, where she teaches astronomy and physics.

Table 12.1 List of TNOs discovered through IASC

TNO	Discovery date	School
2012 HH2	04-19-12	
2014 FP43	03-20-14	
2014 GE45	04-06-14	
2014QF433	08-26-14	National Dali Senior High School (Taiwan)
2014 UH192	10-28-14	
2014 WT69	11-17-14	Hofenfels-Gymnasium Zweibruken (Germany)
2015 FP36	03-19-15	

In 2012, her former junior high school students were in their senior astronomy class when the MPC announced that their 2008 MBA discovery had been numbered and cataloged by the IAU (Paris). Her students named their discovery "Madisonvillehigh" after their school. Orbiting between the Mars and Jupiter at an average distance of 2.89 AU, this asteroid will forever bear the name in honor of the school.

In 2009, at a teacher training workshop, Steven Kirby, a physics teacher at Ranger High School (Ranger, Texas) co-discovered the potentially hazardous asteroid 2009 BD81. This NEO comes within 0.0456 AU of Earth's orbit.

Students have also discovered trans-Neptunian objects, typically located out in the Kuiper Belt at a distance of 40 AU. Table 12.1 is a list of those discoveries to date.

Benefit Derived from Using Remote Observatory Facilities

IASC uses a number of remote observatory facilities. The primary purpose of these facilities is to take confirmation images and measurements of asteroids recently discovered. After the original discovery is made by a student, the MPC requires a confirmation image and measurement within 7–10 days. If this confirmation is not made, the original discovery is considered to be lost, and the discovery process must begin again.

Facility Equipment Used

The following are the facilities currently used by IASC to take the confirmation images:

- 0.81-m RC Mount Lemmon SkyCenter (Sierra Stars Observatory Network)
- 1.3-m RCT Kitt Peak National Observatory (Western Kentucky University)
- 2-m Faulkes Las Cumbras Observatory (Haleakala, Hawaii; Siding Spring, Australia)
- 0.81-m RC Tarleton State University (Texas)

- 0.81-m G.V. Shiaparelli Observatory (Italy)
- 1-m Las Cumbras Observatory (Sutherland, South Africa)
- 1.3-m Astronomical Research Institute (Illinois)
- 2.4-m Magdalena Ridge Observatory (New Mexico Institute of Mining & Technology)

These are all handled remotely with requests for confirmations sent by IASC to these participating sited.

About Patrick Miller

Dr. James Patrick Miller studied at the University of New Mexico, University of California at Los Angeles, and James Cook University (Australia). Since 2005, he has been a Professor of Mathematics at Hardin-Simmons University in Texas. Previously, he served as Dean of Science & Mathematics at Brookhaven College and an Adjunct Professor of Mathematics at the University of Texas at Dallas.

Patrick was President of the National Association of Rocketry for 16 years and subsequently a member of the Board of Trustees until 2000. He was also a member of the Pyrotechnics Committee of the National Fire Protection Association for nearly 25 years, chairing the Rocket Subcommittee.

Patrick has been a guest researcher at the Lawrence Berkeley National Laboratory and a Ph.D. advisor in astronomy at the University of Southern Queensland (Australia). He founded the International Astronomical Search Collaboration in 2006, which now includes 500 schools in more than 65 countries.

University of Iowa Remote Spectroscopy Project (Robert Mutel)

Objective and Goal of Project

The goal of observational astronomy is to extract as much information as possible about the physical conditions of the object being studied. Although beautiful multi-colored images of planets, nebulae, and galaxies dominate the popular imagination and media, spectroscopic observations, in which the light is separated into hundreds or thousands of spectral channels, dominate observing schedules on the world's largest telescopes. This is because spectroscopy allows astronomers to determine a large number of physical conditions that are either difficult or impossible to discern from an image. These include elemental composition, temperature, density, motion (both translational and rotational), and even magnetic field strength.

In spite of these advantages, relatively few amateur astronomers have the capability to conduct spectroscopic observations. This is because conventional spec-trometers are expensive to build, very sensitive to vibration and temperature variations, and difficult to operate because the target object must be placed at the spectrometer entrance with arc-second accuracy. In addition, the spectrometer sub-system often interferes with charge-coupled device (CCD) camera imaging or eyepiece observing, necessitating cumbersome, time-consuming equipment removal and replacement when the spectroscope is installed.

An alternative scheme that offers a low-cost introduction to spectrometry is the transmission grating spectrometer (TGS). A TGS consists of one or more relatively inexpensive gratings inserted before the CCD imaging camera, typically in an empty slot in a filter wheel. The TGS system is a relatively low-resolution spec-trometer, having a typical spectral resolution of order 1–2 nm at visual wavelengths, equivalent to an R value ($=\lambda/\Delta\lambda$) of a few hundred. This is adequate to detect many fundamental astrophysical quantities, such as absorption lines from main-sequence stars; emission lines from planetary nebulae, supernovae, and very hot stars; and even red-shifted lines from bright quasars. However, the spectral resolution is not sufficient for other spectroscopic science targets, e.g., detecting exoplanets using stellar reflex motion or Balmer emission cores from active stars.

In this section, I discuss some technical considerations for installing and using a TGS system in a modest-size telescope and show some examples of TGS spectra taken with the University of Iowa's Rigel telescope, a 37-cm diameter f/14 robotic telescope in southern Arizona. The goal of the Rigel TGS system was to enable a large number of astronomy undergraduates to obtain stellar spectra using the same robotic telescope facility they already had been using for imaging.

How Transmission Gratings Work

A TGS system consists of one or more high-efficiency transmission gratings mounted in front of a CCD imaging camera, typically in place of broadband filters in a filter wheel. As light passes through the grating, it is diffracted into wavelength-dependent angles, as shown in Fig. 12.4.

The grating consists of parallel lines etched in glass, typically several hundred per millimeter. As light passes through the grating, the lines act like independent "Huygens wavelet" sources, causing constructive and destructive interference, depending on diffraction angle. For constructive interference, the path difference between adjacent Huygens wave sources must be an integer multiple of a wave-length. Hence, for the first-order ($n=1$), the relationship between wavelength and the diffracted angle is:

$$\sin\theta = \frac{\lambda}{a} \qquad (12.1)$$

where λ is the wavelength and a is the line separation. For example, a grating with 600 lines per millimeter (lpmm) has a line separation $a=1/600$ mm $=1.67$ μm (μm $=10^{-6}$ m), and hence $\theta=17$ degrees at $\lambda=500$ nm. Longer wavelengths are dif-fracted to larger angles and shorter wavelengths to smaller angles, so the incoming

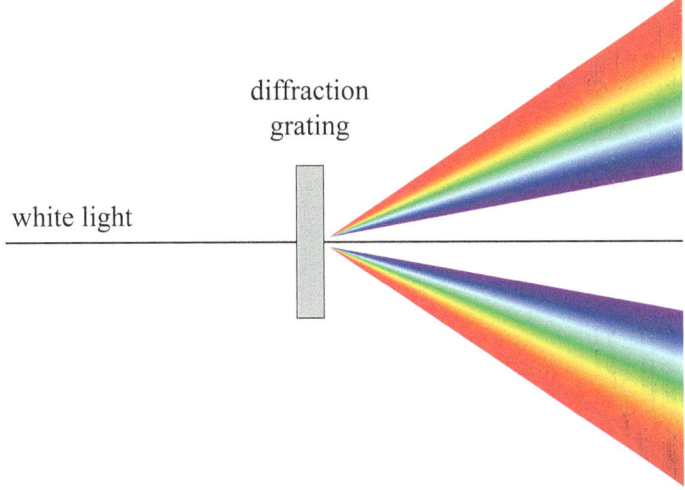

Fig. 12.4 Diffraction grating

light is spread across the detector in a wavelength-dependent manner. This is the key feature of all spectrometers. The converging light path intercepts the grating at the filter wheel, which is ahead of the focal plane, where the CCD camera is located.

Grating Efficiency

Figure 12.4 shows that the diffracted rays propagate in both directions with equal intensity. However, modern gratings are "blazed," which improves efficiency on one side at the expense of the other. The grating efficiency is a function of the line density, becoming lower with smaller line spacings. The gratings used on the Rigel TGS system were purchased from Thorlabs; the efficiency versus wavelength is shown in Fig. 12.5. The gratings are marked with a small arrow on the edge indicating the direction of maximum response.

Focal Plane Plate Scale

The focal length (*FL*) of a telescope is the product of the primary mirror diameter *D* and the *f*-ratio, $FL = Df$. The plate scale ζ, expressed in microns per microradian, is equal to the *FL* divided by 1 rad. For the Rigel telescope, the *f*-ratio is 14, so $FL = 37 \text{ cm} \times 14 = 5.18$ m. Hence the plate scale $\zeta = 25$ µm/arc-second. Assuming mean seeing is 2 arc-seconds, a stellar image will have a linear dimension 50 µm on the CCD sensor. This dimension determines the spectral resolution of a TGS system for point sources (stars) as described in the next section.

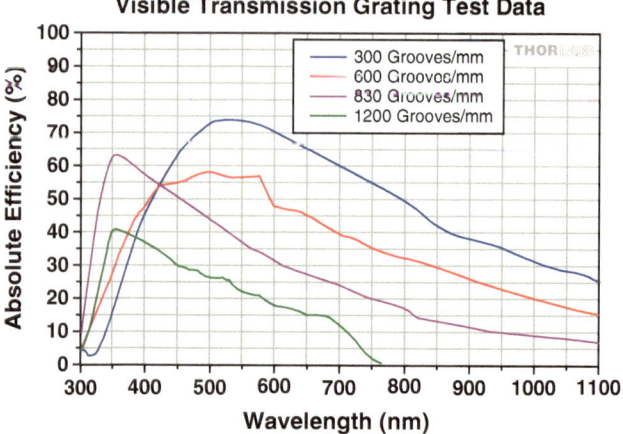

Fig. 12.5 Transmission grating efficiency

Spectral Resolution

A conventional spectrometer has a narrow (typically 25–50 μm) slit in front of the diffraction grating or prism that determines the spectral resolution. By contrast, a TGS system is typically configured without a slit. In this case, the spectral resolution depends on the telescope's seeing, the grating density, the grating-detector separation, and the wavelength. Figure 12.6 shows the geometry of the grating and its relationship to the focal plane in more detail. At a wavelength λ, a star's light is diffracted at a horizontal displacement from the field center according to the following formula:

$$x(\lambda) = z_0 \cdot \tan\left[\sin^{-1}\left(\frac{\lambda}{a}\right)\right] \approx z_0\frac{\lambda}{a} \tag{12.2}$$

By differentiating this equation and equating dx with the linear spread of a point source at the focal plane caused by seeing, the spectra resolution can be written:

$$\lambda = \frac{a}{z_0}\theta_s \cdot FL \tag{12.3}$$

where θ_s is the seeing angular size (in radians).

Defocus Degradation Caused by Curved Focal Plane

The spectral resolution calculated above does not take into account the degrading effect of defocusing caused by the curved focus plane, as illustrated in Fig. 12.7.

Although conventional spectrometer designs introduce curved mirrors to compensate for the curved focal plane of the diffracted light, the simplified TGS design

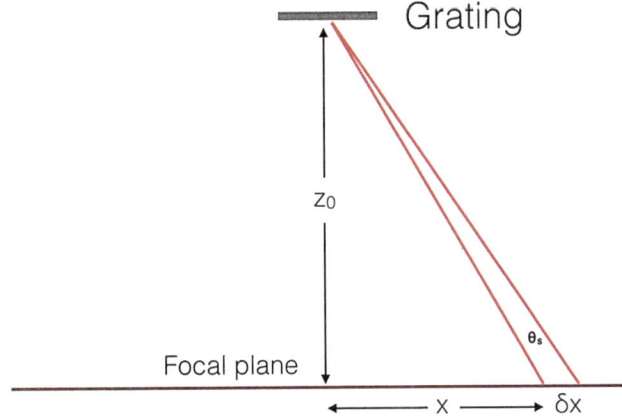

Fig. 12.6 Focal plane geometry

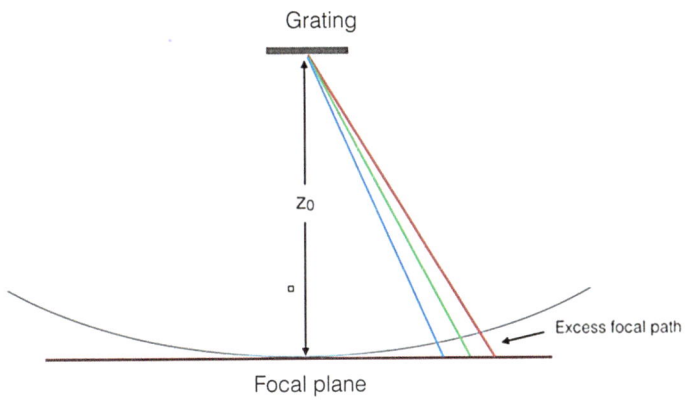

Fig. 12.7 Excess path length

uses the imaging camera with a flat focal plane and no curvature compensation. For TGS observations, the telescope is refocused to compensate for the excess path length at a chosen wavelength. The additional focal length as a function of wavelength is:

$$z = \frac{z_0}{2} \cdot \left(\frac{\lambda}{a} \right)^2 \qquad (12.4)$$

For example, the Rigel telescope with a 600-lpmm grating has an excess path length $\Delta z \approx 1$–4 mm between 350 nm and 750 nm. At the wavelength corresponding to the new focus, the spectral resolution is given by the seeing Eq. (12.3). However, because the excess path length is wavelength dependent, at other wavelengths, the image is defocused, so the spectral resolution is degraded. The degradation becomes significant when the differential excess path length becomes comparable to the depth of focus. The depth of focus is given by:

$$d_f = \theta_s \cdot D \cdot f^2 \tag{12.5}$$

where θ_s is the seeing disk, D is the telescope diameter, and f is the focal ratio. For the Rigel telescope, this is about 200 μm for 2.0 arc-seconds seeing. The condition that the differential path length does not exceed one-half the depth of focus results in an effective full-width wavelength range of:

$$\Delta\lambda = \frac{1}{z_0 \cdot \theta_s \cdot D}\left(\frac{a}{\lambda \cdot f}\right)^2 \tag{12.6}$$

This equation is plotted in Fig. 12.8 for two sample telescopes: the Rigel telescope (37 cm, f/14), and a 20-cm diameter telescope with f/8 optics (e.g., Celestron C8). For each telescope, both 300-lpmm and 600-lpmm gratings are plotted. The largest in-focus wavelength range occurs for longer focal ratios and lower resolution gratings.

The effect of focus position on spectral resolution is illustrated in Fig. 12.9, which shows Rigel spectra of the star γ Cas at focus positions between 13,900 μm and 14,150 μm.

Observing Considerations Using TGS Systems

Signal-to-noise ratio (SNR). The SNR of a photon-counting detector is proportional to the square root of the product of exposure time and bandwidth. Because the spectrum is spread over many resolution channels, each channel's SNR is lower than the SNR of a broadband filter by the square of the ratio of the bandwidths. For example, comparing a spectral resolution of 2 nm with an unfiltered image that has an effective bandwidth 500 nm, the SNR is reduced by a factor $\sqrt{(2/500)} \approx 0.06$ (6 %) compared with the unfiltered SNR. In addition, the grating efficiency (discussed in Section "Grating Efficiency") also reduces the light throughput. The fractional reduction depends on the grating and wavelength but varies from 0.2 to 0.6, so the overall channel SNR is typically 1–2 % of the unfiltered star's SNR using the same exposure time.

Target image displacement and CCD sensor size. Because the grating blaze favors one side of the dispersed spectrum, it is important to position the target on the correct side so the brighter first-order spectrum is displayed on the CCD sensor. In addition, the CCD sensor must be large enough that the entire visible spectrum fits on the sensor (unless the observer is only interested in a spectral sub-region).

The wavelength-dependent displacement of the spectrum from the zeroth-order image is given by Eq. (12.2). For the Rigel telescope and the 600-lpmm grating, this

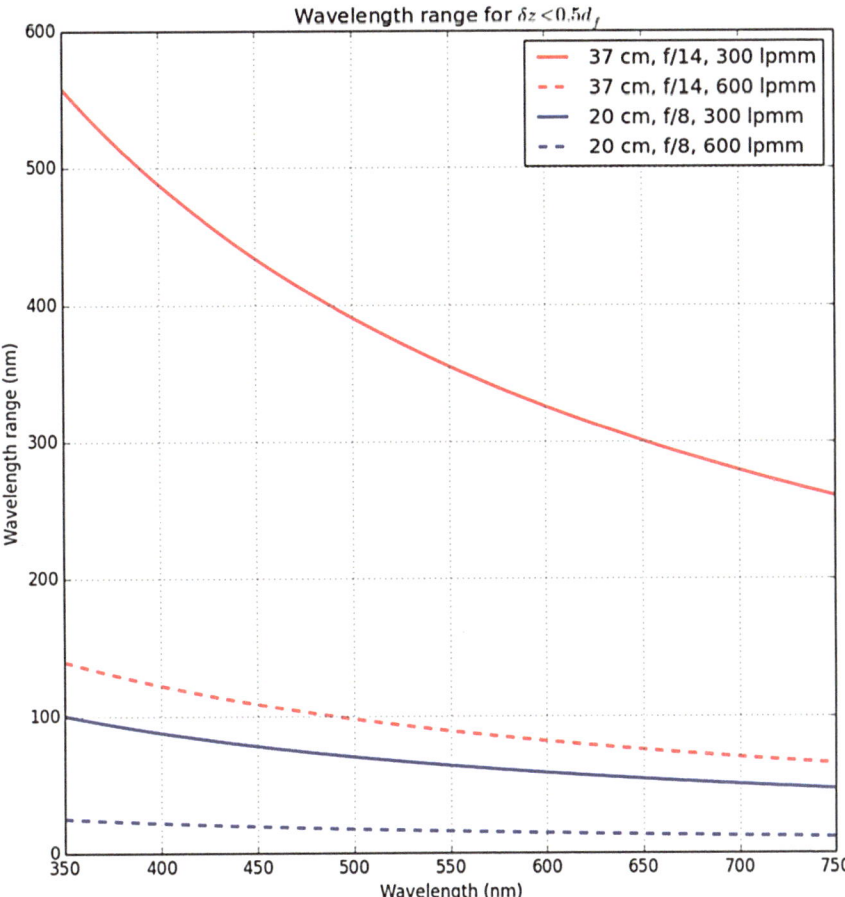

Fig. 12.8 In-focus distance versus wavelength for two telescopes

ranges from 9 mm (350 nm) to 20 mm (750 nm). The CCD sensor is 37 mm × 37 mm, so the zeroth-order star image and the entire visible spectrum fits comfortably on the sensor. A sample spectral image is shown in Fig. 12.10 (Vega, cropped to show only the strip spectrum and zeroth-order star at left). The hydrogen Balmer lines are evident, and the out-of-focus zero-order image of the star is at the left.

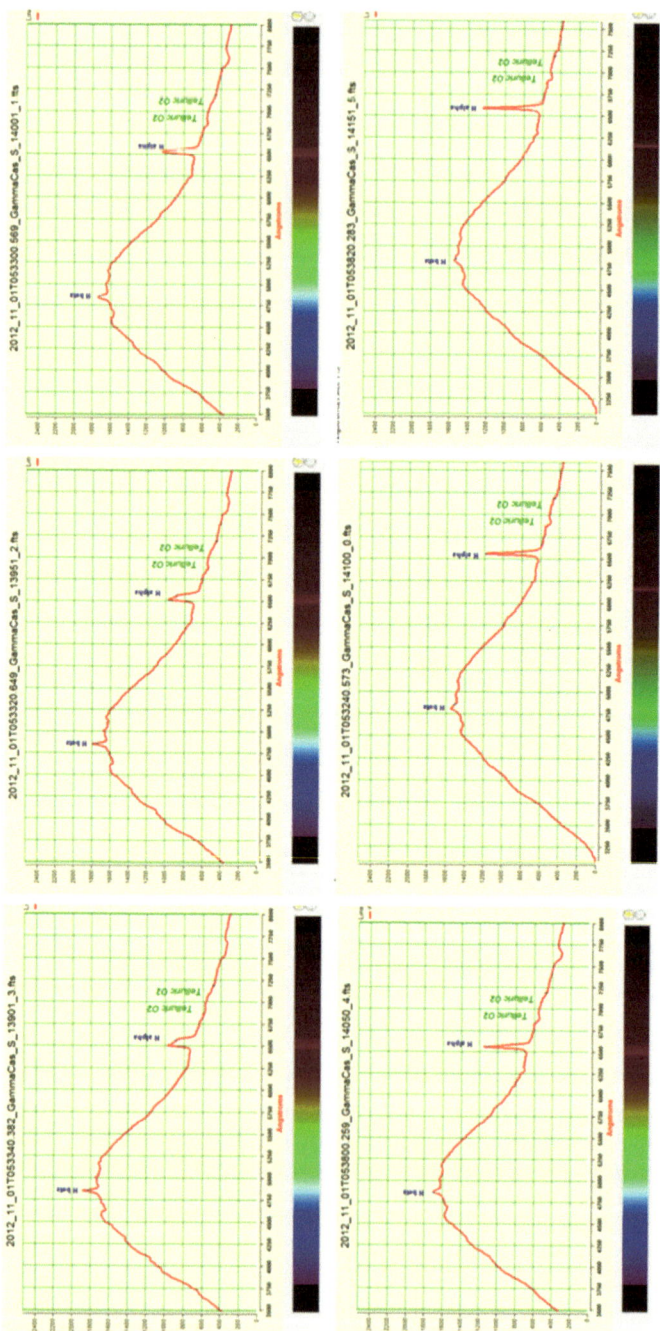

Fig. 12.9 3×2 Gamma Cas spectra showing defocusing

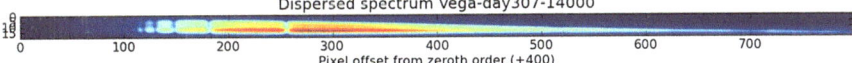

Fig. 12.10 Vega strip spectrum

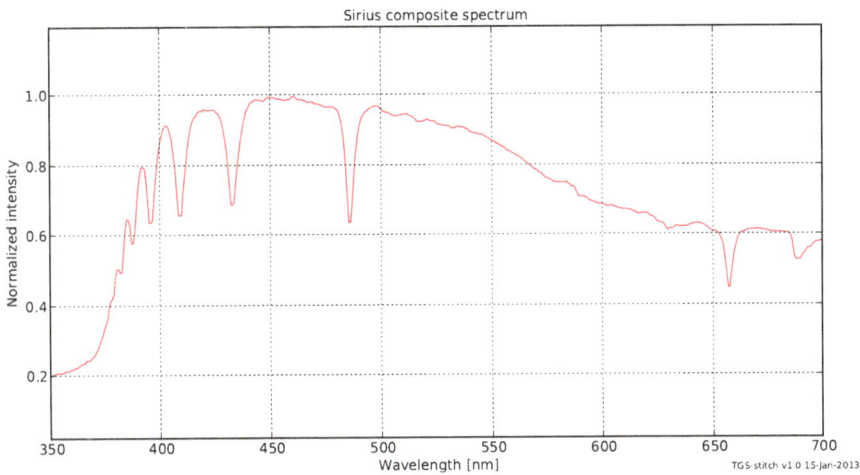

Fig. 12.11 Sirius composite spectrum

Results

The example spectra shown in Figs. 12.11, 12.12, 12.13, and 12.14 were all taken using the Rigel telescope and a 600-lpmm grating. Most are composite spectra taken at several focus settings and "stitched together" to make a composite spectrum.

Benefit Derived from Using Remote Observatory Facilities

The most obvious benefit of using remote observatory facilities for transmission grating spectroscopic observations is the ability to take spectra robotically without the need to manually position the object on a slit. The resulting spectra, although of modest spectral resolution, enable the observer equipped with a small telescope and a conventional filter wheel to obtain spectra that can be used to distinguish stars of different spectral types, identify emission lines in emission nebulae and emission-line stars (e.g., Wolf-Rayed and Be stars), and even determine the redshift of bright quasars. This convenience comes with several limitations, e.g., modest spectral reso-

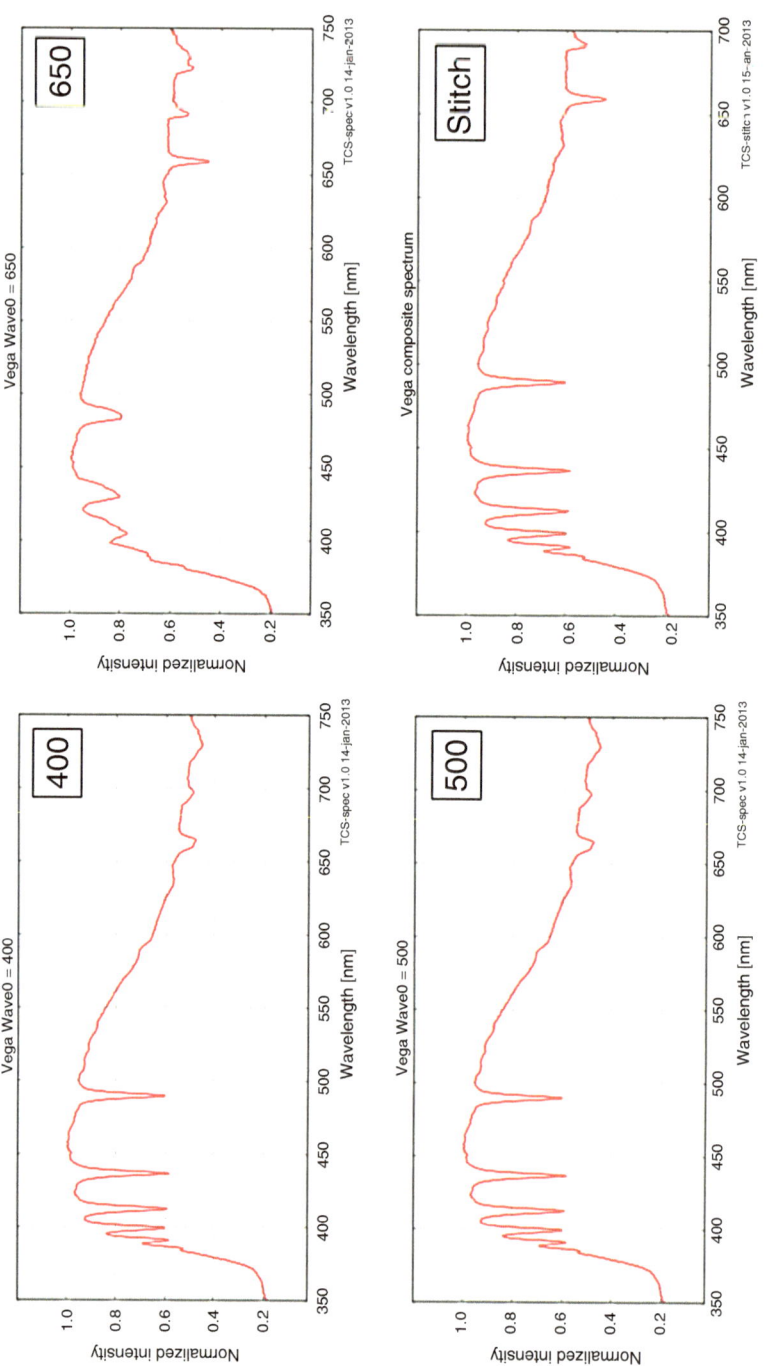

Fig. 12.12 Four stellar spectra

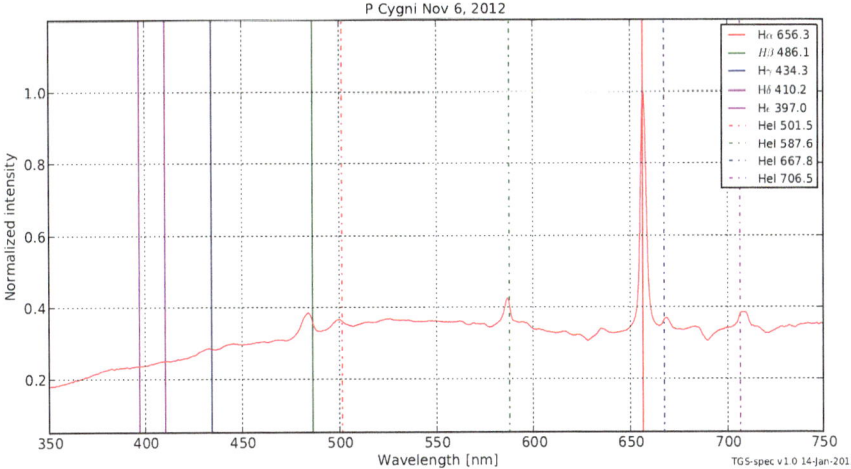

Fig. 12.13 P Cygni spectrum

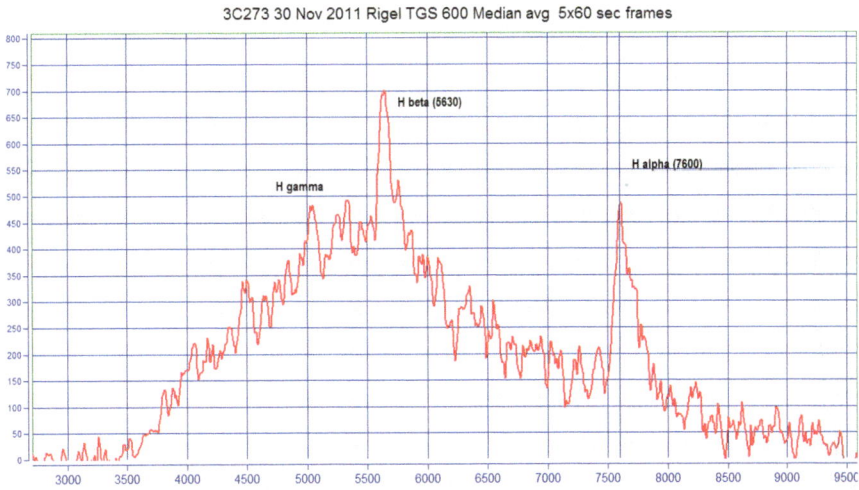

Fig. 12.14 3C273 spectrum

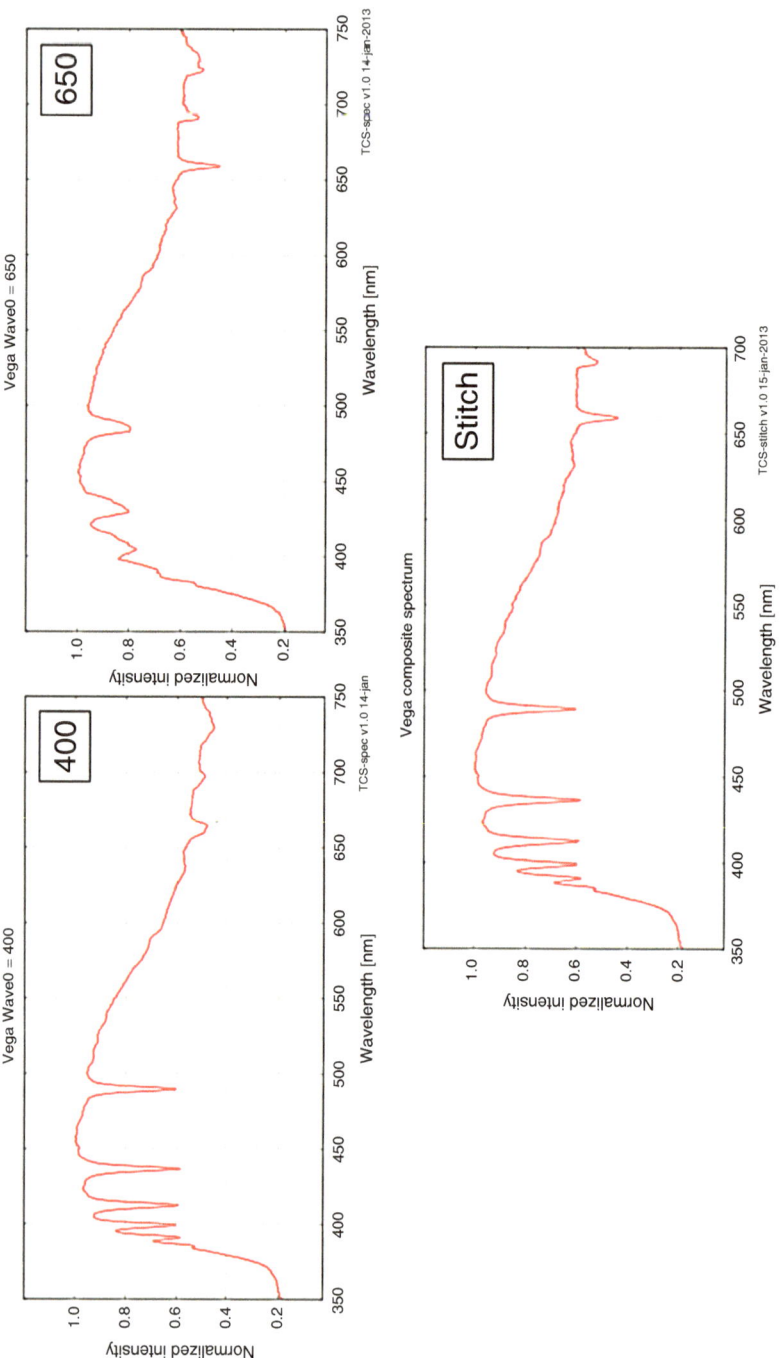

Fig. 12.15 Stitched Vega spectrum—three panels

lution, the need for specialized software to analyze the spectra, and possibly contamination of the target spectrum by light from stars overlaid on the target spectrum.

Spectral Analysis Software. The spectra obtained using TGS systems are strip recordings on the CCD image (e.g., Fig. 12.15). To turn these into calibration one-dimensional plots of intensity versus wavelength, specialized software is needed. Two such software packages include RSpec® and TGS-spec, both described below. In addition, to mitigate the effects of focal plane defocusing, the program TGS-stitch was developed to combine several spectra taken with multiple focus positions.

1. RSpec®. This is a Microsoft® Windows-based commercial program that is well-suited to reducing TGS spectra. It includes a sophisticated polynomial wavelength calibration function and the ability to analyze line spectra that are slightly misaligned with the CCD frame. The reader is encouraged to study the Rspec® website (www.rspec-astro.com) for complete details.
2. TGS-spec. This is a free Python-based program that was developed at the University of Iowa. It runs on Macintosh®, Linux, and Microsoft® Windows computers and has a user-friendly graphical user interface. It reads standard FITS-format images. Complete details and instructions for downloading and installing are at http://astro.physics.uiowa.edu/rigel/spectroscopy. Most of the spectra shown in this section were prepared using TGS-spec.
3. TGS-stitch. To observe a star's spectrum over the full visible range (350–700 nm), either a grating-telescope system should be chosen that has a spectral bandwidth sufficient to cover the entire range without refocusing, or several spectra should be taken at different focus positions and the resulting spectra stitched together to produce a complete spectrum. The former is possible with low-resolution gratings (e.g., 300 lpmm) and the Rigel telescope, but the limiting resolution is relatively low, about 2 nm. Alternatively, the stitch technique uses individual spectra at differing focal offsets and stitches the spectra together to form a composite spectrum. An example of this technique is shown in Fig. 12.15 for the star Vega.

Facility and Equipment Used

The Rigel telescope (Fig. 12.16) consists of an f/14 classical Cassegrain optical tube assembly (OTA) with a 37-cm- (14.5-in.-) diameter primary mirror mounted on an equatorial mount. The OTA and mount were manufactured by Optical Mechanics Inc. in Iowa City, IA. The telescope is located at the Winer Observatory (winer.org) near Sonoita, AZ. The observatory houses several University-based robotic telescopes and has a large roll-off roof, weather monitoring equipment, and an all-sky camera.

The CCD camera is a Finger Lakes Instruments® (FLI) Proline PL-16803 (4 K×4 K, 12 μm pixels, front-illuminated). The filter wheel is a FLI Centerline 10-position wheel using 50 mm×50 mm square filters. The transmission gratings were purchased from Thorlabs (www.thorlabs.com), model GT50-03 (300 lpmm)

Fig. 12.16 Rigel telescope

and GT50-06 (600 lpmm). The geometry of the camera, filter wheel, and grating is shown in Fig. 12.17.

About Robert Mutel

Dr. Robert Mutel is a Professor of Astronomy at the University of Iowa in Iowa City, IA. He received his Ph.D. at the University of Colorado in 1975. He is recognized as one of the leading experts in radio astronomy. His primary interests in this field are the study of active and early stars and space physics, especially magnetospheric radio emission processes.

Robert was one of the pioneers in developing robotic/remote telescope systems for teaching and research projects. He and his University of Iowa students designed and built an automated 0.5-m robotic telescope called the Iowa Robotic Observatory. In 1998, they moved the telescope to the Winer Observatory in Sonoita, AZ, and operated the telescope remotely from Iowa. Shortly afterward, Robert received a

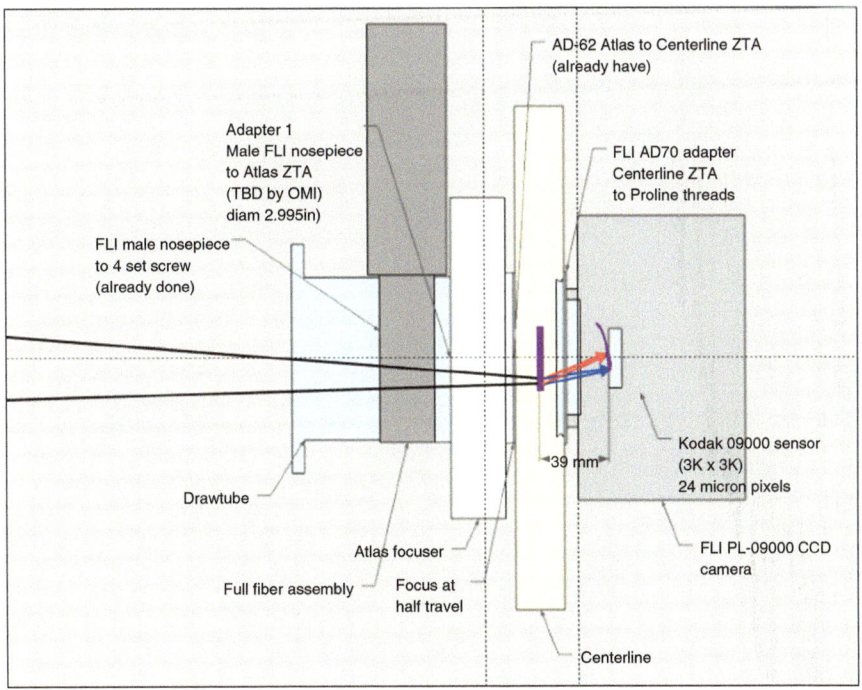

Fig. 12.17 Rigel Filter wheel/camera light path

grant to build a new, more advanced robotic 14.5-in. (0.35-m) telescope designed and built by James Mulherin, founder of Torus Optics. This remote telescope, named Rigel, was the foundation for the University of Arizona's remote astronomy laboratories and research projects.

In 2008 Robert and the University of Iowa partnered with Rich Williams (one of the authors of this book) to join the Sierra Stars Observatory Network (SSON). The idea was to use the unused and otherwise wasted time in the nightly observing run queues and add SSON schedule requests to take advantage of the unused resource. SSON paid for the resulting images delivered to SSON, which provided a recurring revenue stream to help offset the cost of housing and operating the telescope at the Winer Observatory. The Rigel telescope TGS was the first commercial remote spectrograph available for use by the general public through SSON.

In 2013, Robert received a Carver Trust grant to fund a new remote telescope at the Winer Observatory called the Van Allen Gemini Telescope Project. The project's telescope is a 20-in. (0.5-m) CDK telescope manufactured by PlaneWave with a CCD camera containing a very high-quantum efficiency back-illuminated CCD chip for imaging. In addition, the telescope will have a high-resolution fiber-fed spectrograph, which will give students and researchers the ability to do interesting cutting-edge spectroscopy projects. The telescope begins operation in the fall semester of 2015.

Chapter 13

Esthetic Imaging Projects

The previous chapters in Part III highlight case studies of astronomy science projects. Many people are also interested in taking images of astronomical objects to show their beauty. Remote astronomy observatories enable them to acquire high-quality images from locations with dark skies and good seeing.

This chapter highlights esthetic imaging projects from the perspective of someone just starting out and from someone who is recognized as one of the greatest astro-imagers in the world today. The figures are courtesy of the respective contributor.

First Experiences Using Remote Observatories for Aesthetic Imaging (David Galbraith)

As an amateur astronomer who has been interested in both photography and physics since childhood, I have enjoyed experimenting with different means to capture astrophotographs. I completed my Ph.D. in wildlife biology from Queen's University at Kingston and now serve as the Head of Science at Royal Botanical Gardens (in Hamilton and Burlington, Ontario), where I occasionally lead public programs in astronomy and night sky photography, but otherwise I don't have any professional involvement in astronomy. As a member of local amateur astronomy clubs, I have enjoyed public demonstrations of night sky observing and have undertaken quite a few night sky photography projects for purely aesthetic purposes. In 2011, I started to get a little more serious about the hobby of amateur astronomy

© Springer International Publishing Switzerland 2015
G.R. Hubbell et al., *Remote Observatories for Amateur Astronomers*, The Patrick
Moore Practical Astronomy Series, DOI 10.1007/978-3-319-21906-6_13

and began to assemble some of my own equipment. At the same time, I became aware of opportunities to undertake remote observations that had not been discussed in the club meetings I attended.

I started using some remote observatories in 2013, focusing on facilities offered by the Sierra Stars Observatory Network (SSON) (Galbraith 2014a, b). The ability to access professional-grade instruments on-line was very exciting and provided an affordable and easy way to take images that were of high quality.

Objective and Goal of Project

In honor of International Astronomy Day in 2013, I decided to capture a color image of the Whirlpool Galaxy (NGC 5194 (Messier 51A)) on that day. The Whirlpool Galaxy is a large, attractive spiral galaxy with a companion barred elliptical galaxy (NGC5195 (Messier 51B)) in the northern hemisphere constellation Canes Venatici. Sitting at a distance of approximately 23 million light years from Earth, we see this classic galaxy nearly disc-on. With a long dimension of approximately 11 arc-minutes, it spans approximately one third the angular size of the full Moon, at an apparent magnitude of 8.4 (Frommert and Kronberg 2013).

I selected the SSON 61-cm Cassegrain telescope in southern California for my project. To use the telescope from my home in Hamilton, Ontario, Canada, I programmed it using the SSON web interface (http://www.sierrastars.com). This was my first project in LRGB imaging.

To capture this large and fairly bright object in LRGB color in a small number of sessions, I used the SSON telescope to capture multiple images overnight on April 19–20, 2013. Each of the four layers needed to construct the final color image (luminance was captured without a filter; the other three layers were captured with the available red, blue, and green colored filters) was taken as three separate 300-s exposures. The resulting monochrome images were stacked and then combined using Fitswork software (http://www.fitswork.de/software/softw_en.php) and adjusted with a free copy of Adobe® Photoshop® Elements. I did not have access to Adobe® Photoshop®, which is the image management software commonly used by amateurs for refining astronomic images.

Results

The exposures prepared for this project produced FITS files with embedded images 1528×1528 pixels in size, as a result of preset 2 by 2 binning in the camera (Fig. 13.1). The 300-s exposures produced an overexposure in the core of M51A but did allow the extended and fainter operations of the galaxy to be imaged. Inspection of the images suggests that the SSON combination of telescope and

Fig. 13.1 An uncropped 1528 × 1528 pixel image from the April 2013 imaging run of M51 captured by the 61-cm SSON telescope

camera used for this project covers about 25 arc-minutes, adequately capturing the faint, extended portions of the nebula. This exposure used the V filter setting and has only been adjusted using ESO/ESA/NASA FITS Liberator software's histogram function to give an overall impression of the scale and quality of the image. The Moon was at 91 % full and about 112 degrees away from the galaxy. The image recording software at the observatory automatically applies corrections to each frame for bias, thermal artifacts, flat field corrections, and bad columns.

Aligning and stacking the images as a monochrome rendering with freely available software revealed considerable detail in the dust lanes and other areas of the Whirlpool not clear on any single frame (Fig. 13.2). Details in the dust lanes of M51A became much more apparent, and reduction of noise in the surrounding sky is apparent compared with any single frame, such as that presented as Fig. 13.1.

Reintegration of the frames exposed using color filters produced a pleasing if somewhat lower-resolution LRGB image (Fig. 13.3).

Fig. 13.2 A stacked monochrome image produced by using all 12 exposures captured for this project

Benefit Derived from Using Remote Observatory Facilities

The project resulted in an acceptable, if not particularly high-resolution, color image of M51A and M51B for International Astronomy Day in 2013.

The primary benefits of using the remote observatory facility compared with a home telescope for a project like this were convenience and time-saving. Because the observatory is already set up and professionally maintained in a dark sky area, the user does not have to undertake travel, set up, or alignment. Even once set up, the exposure time and guiding necessary to produce each long exposure (in this case, 300 s each) is replaced by a few seconds of input to the simple web-based user interface provided by SSON. The image produced in this project was actually limited somewhat in quality and resolution by the author's early learner stage in LRGB imaging and the free software employed for stacking and image improvement.

Fig. 13.3 Finished LRGB color image for International Astronomy Day, 2013

Facility Equipment Used

The telescope selected was the SSON 61 cm f/10 Optical Mechanics Nighthawk CC06 Cassegrain in southern California. The telescope was fitted with a Finger Lakes Instrumentation FLI ProLine PL09000 camera.

About David Galbraith

David Galbraith's interest in astronomy was kindled in childhood by spending many clear, dark nights along the Canadian shore of Lake Huron with his family, using a 4.5-in. Newtonian reflector to peer at the Moon and planets. Although he pursued biological science as a career, David has never lost his fascination with physics and astronomy. He serves as Head of Science at Royal Botanical Gardens (Hamilton and Burlington, Ontario, Canada) where he has led public night-time and solar observing events. He is occasionally active in both the Hamilton Amateur Astronomers and the Royal Astronomical Society of Canada—Hamilton Centre.

Remote Esthetic Imaging Using the Mount Lemmon SkyCenter Telescopes (Adam Block)

Objective and Goal of the Project

I am an astronomer and the observatory manager of the University of Arizona's Mount Lemmon SkyCenter located near Tucson, Arizona. I run regular public outreach programs at the summit to educate people about astronomy and telescopes. I also teach hands-on courses on astronomical imaging using the Sky telescopes locally at the summit.

When the SkyCenter joined the Sierra Stars Observatory Network (SSON), users began taking images remotely to create their own beautiful images of galaxies and nebulae. This has proven to be a very effective way for people to achieve remarkable results remotely and in an affordable way without having to travel to Arizona to use the observatory telescopes. Now anyone in the world can access our 32-in. (0.8-m) and 24-in. (0.6-m) telescopes to schedule images to be taken using the SSON web-based scheduling system.

Results

SSON users who schedule images for their esthetic imaging projects receive their image data fully calibrated and ready to combine and process as they see fit. They use image processing software to create their final product. The most important part of creating a beautiful image of an interesting object is to start with the highest quality image data. While image processing can bring out and highlight details, if the data you start out with are not great, your final result will suffer.

The SkyCenter observatories on Mount Lemmon are 9157 ft (2791 m) above sea level. The sky is very dark, and the seeing often has 1 arc-second or better resolution. It is one of the very best observing locations in the southwestern United States.

The images shown in Figs. 13.4, 13.5, and 13.6 demonstrate the fantastic results you can achieve with image data using the Mount Lemmon SkyCenter observatories.

Fig. 13.4 Galaxy NGC 1398 from image data taken with the 32-in. (0.8-m) Schulman telescope
Credit: Adam Block

Benefit Derived from Using Remote Observatory Facilities

Traveling to the summit of a mountain to acquire images locally is not an option for most people. And even when it is, it can be expensive and time consuming. Although I live locally in Tucson and drive up Mount Lemmon to the observatories regularly, I frequently operate the observatories remotely from my home or even on the road as far away as Japan. Fortunately for me, a professional technical staff oversees and maintains the Mount Lemmon SkyCenter and adjacent observatories on the summit.

The SkyCenter's partnership with SSON makes these high-precision professional telescopes available to anyone who wants to use them for their own esthetic imaging projects. Because the telescopes are relatively large and have sensitive, high quantum efficiency CCD cameras, you can achieve fantastic results even with relatively short total exposure times. It's easy to acquire your image data remotely and the cost is reasonable.

Fig. 13.5 Cepheus Bubble nebula Sh2-140 from image data taken with the 32-in. (0.8-m) Schulman telescope Credit: Adam Block

Facility Equipment Used

I use the Mount Lemmon SkyCenter's 24-in. Philips (0.61-m) telescope and the 32-in. (0.81-m) telescope for astro-imaging projects.

About Adam Block

Adam Block's life-long goal to be an astronomer began at the early age of 4. He specifically selected the University of Arizona to continue his pursuits and graduated with a B.S. in Astronomy and Physics in 1996. Adam spent the next 9 years developing and administering public outreach programs at the National Observatory

Fig. 13.6 Helix Nebula (NGC 7293) from image data taken with the 32-in. (0.8-m) Schulman telescope Credit: Adam Block

(Kitt Peak). Since 2007, his dream to create the foremost public outreach experiences in astronomy is being realized with a return to the university through the Steward Observatory and the College of Science at the Mount Lemmon SkyCenter.

Adam is most well-known for his abilities to speak and communicate difficult astronomy concepts using straightforward, creative methods. Over the past 15+ years, he has hosted many thousands of evening sessions for the public. He strives to maintain quality programs that are fresh and exciting with unflagging enthusiasm.

Adam is also recognized around the world as a leading astrophotographer. The images he produces as part of public outreach programs are published in magazines, books, posters, and widely on the Internet. His images have graced NASA's "Astronomy Picture of the Day" website more than 70 times. Both amateur and professional astronomers use these images as standard references of quality and precision. In 2012, Adam received the prestigious "Hubble Award" from the Advanced Imaging Conference—one of the highest awards for work in astrophotography recognized around the world. In 2013, the Greenwich Royal Observatory honored him as astrophotographer of the year for best deep space astrophotography. This year, Adam began writing monthly columns in *Astronomy* magazine about image processing.

References

Galbraith DA (2014a) Trying remote astronomy with the Sierra Stars Observatory Networks. *Event Horizon* 21(3): 7-10. On-line document. URL: http://www.amateurastronomy.org/wp-content/uploads/2014/09/January2014.pdf accessed 7 May 2015

Galbraith DA (2014b) Publicly Accessible Robotic Telescope Networks. *Event Horizon*, the Newsletter of the Hamilton Amateur Astronomers 21(4), 7–11. On-line document. URL: http://www.amateurastronomy.org/wp-content/uploads/2014/09/February2014.pdf accessed 7 May 2015

Frommert H, Kronberg C (2013) Messier 51. On-line document. URL: http://messier.seds.org/m/m051.html accessed 19 April 2015

Index

© Springer International Publishing Switzerland 2015   227
G.R. Hubbell et al., *Remote Observatories for Amateur Astronomers*, The Patrick
Moore Practical Astronomy Series, DOI 10.1007/978-3-319-21906-6